Agricultural and Hydrological Applications of Remote Sensing

Agricultural and Hydrological Applications of Remote Sensing

Brindra Chauhan

RANDOM

RANDOM PUBLICATIONS

NEW DELHI (INDIA)

Agricultural and Hydrological Applications of Remote Sensing

ISBN 978-93-5111-600-4

Published in 2015 in India by

RANDOM PUBLICATIONS

Reprint, 2017

4376-A/4B, Gali Murari Lal, Ansari Road
New Delhi-110 002
Phone : +9111-43580356, 011-23289044, 011-43142548
e-mail: sales@randompublications.com,
info@randompublications.com, randomexports@gmail.com

Type Setting by : Friends Media, Delhi-110089
Digitally Printed at : Replika Press Pvt. Ltd.

Preface

A brief review of research in remote sensing of agricultural and hydrological resources indicates that there are many positive results, and some techniques have been applied operationally. Currently, remote sensing data are being used operationally in precipitation estimates, soil moisture measurements for irrigation scheduling, snow agricultural and hydrological equivalent and snow cover extent assessments, seasonal and short term snowmelt runoff forecasts, and surface agricultural and hydrological inventories. In the next decade other operational applications are likely using remote measurements of land cover, sediment loads, erosion, and hydrological, and a real inputs to hydrological models. Many research challenges remain, and significant progress is expected in areas like albedo measurements, energy budgets, and evapotranspiration estimation. The research in remote sensing and agricultural and hydrological resources also has much relevance for related studies of climate change and global habitability.

I would like to thank my team for standing beside me throughout my career and writing this book. My special thanks go to "Random Publications" who have published the book.

– Brindra Chauhan

Contents

1

Hydrology, Soil and Water Group

THE HYDROLOGICAL CYCLE

An understanding of the hydrological cycle is essential for the effective management of rainwater and soil water. Water occurs not only as a liquid, but also as a solid (*e.g.* hail, snow) and as a gas - water vapour.

The total amount of water in the world is constant, but water is continuously changing from one form to another and is continuously moving at different speeds. These interrelationships are shown in a simplified form at a regional scale.

Heat from the sun causes water at the surface of oceans, lakes and rivers to change into water vapour in a process called evaporation. Transpiration in plants is a similar process, in which water is absorbed from the soil by plant roots and transported up the stem to the leaves, from where it is released (transpired) as water vapour into the atmosphere.

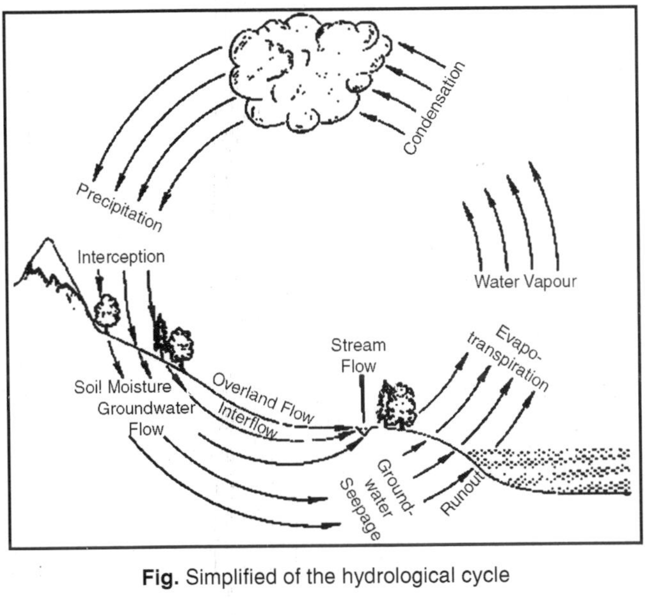

Fig. Simplified of the hydrological cycle

As the water vapour produced by evaporation and transpiration rises into the atmosphere, so the temperature decreases and water vapour changes into water droplets (condensation), which accumulate as clouds. Depending on their size, these may be released as rainfall.

Once rainfall reaches the land surface it can infiltrate into the soil, run off over the surface as overland flow, or accumulate on plant leaves or in puddles from where it evaporates back to the atmosphere. A combination of these processes is commonly the case. The rainfall that infiltrates into the soil forms part of the soil water, of which some may be used by plants for transpiration, some may return to the atmosphere through evaporation from the soil surface, and some - if sufficient infiltration occurs - may move beyond the rooting zone to the groundwater.

The groundwater moves laterally and slowly towards the sea to complete the hydrological cycle, but part of it will seep into springs, streams, rivers and lakes on the way. In this way the groundwater maintains the water level in wells, and the continuity of river and streamflow during dry periods.

Rainwater that runs off the land moves rapidly downhill towards river courses, contributing to peak flows, and is of great concern. Run-off is not only a waste of rainfall that could have contributed to crop production and groundwater supplies, but it frequently causes floods or damage to roads and farmland, and erodes soil that is redeposited in river courses and reservoirs downstream. Groundwater is derived from rainwater that has infiltrated into the soil and drained beyond the rooting zone in excess of both the quantity needed for the crop or the vegetation and the water-storage capacity of the soil.

Groundwater moves very slowly through subsoil materials in the direction of the dominant drainage course. If its upper surface, the water table, does not sink below the level of the streambed, water is released to springs that feed streams and tributaries. This occurs throughout the year and in this way groundwater acts as a buffer in maintaining stream base flows and water levels in wells during dry periods. In soils with relatively impermeable subsoil layers beneath more highly permeable layers, perched water tables may develop above the groundwater, due to water being held up by the impermeable layers. The water in a perched water table, sometimes referred to as interflow, will slowly move laterally and may emerge into stream courses or springs at lower elevations. It does not contribute directly to the groundwater. The presence of groundwater or a perched water table is indicated by saturated soils, and usually by a dominance of light grey, bluish-grey, bluish or greenish colours. These colours are typical of certain iron compounds that only form in waterlogged soils where oxygen is lacking.

The amount of rainfall that percolates beyond the lower limit of the rooting zone towards the groundwater will depend on the amount of water used for transpiration by the crops or vegetation. For a particular climate and soil type, forest transpires more water than grassland, which generally uses more water than crops.

The high water use by forest is due to its generally greater transpiration rate, the longer period of transpiration compared with crops, and the deeper roots enabling it to absorb water from greater depths. Changes in land use can therefore affect the quantity of water transpired and hence the quantity reaching the groundwater. Replacing forest vegetation with grassland or annual crops may increase deep drainage and so provide higher base flows in streams and rivers. Changes in soil management can also affect the quantity of deep drainage replenishing groundwater. The introduction of poor management practices that increase the proportion of rainfall lost as run-off will reduce base flows and increase peak flows and the incidence of flooding. Conversely, an improvement in soil and nutrient management will lead to higher grain and foliage production, higher transpiration rates, and hence less recharge.

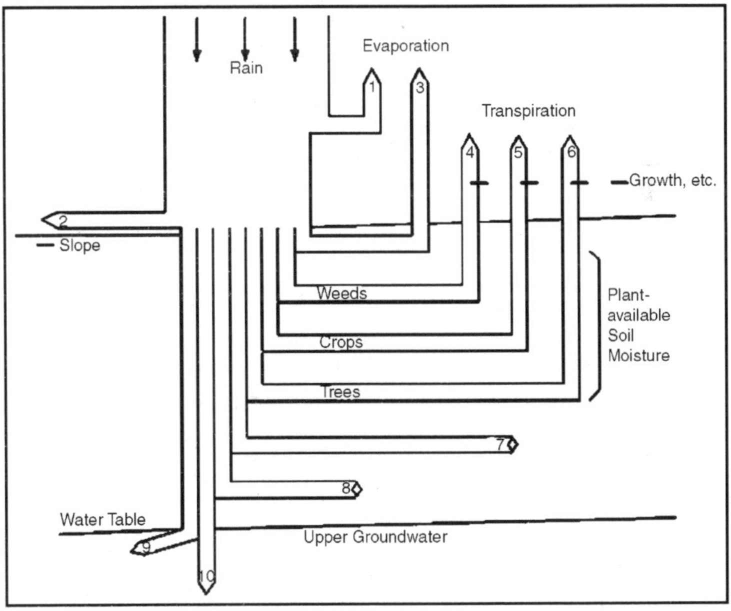

Fig. The sequence of destinations of rainwater

Legend:

- Direct evaporation from wetted leaf surfaces.
- Surface run-off/stormflow.
- Direct evaporation from the soil surface.
- Plant-available soil moisture within root-range of existing weeds, crops, trees.
- Soil moisture within root-range of existing plants but held at tensions unavailable to them.

- Soil moistures held at all tensions, but below root-depth of existing plants.
- Water not captured by roots and small pores, moving to groundwater and streamflow.
- Leakage to deep groundwater beneath catchment floor.

In order to consider rainwater for plants and for groundwater as parts of a sequence, it is important to have a mental picture of its journey. After passing through the atmosphere in response to gravity, water from rain or irrigation travels to some or all of the following destinations.

Management of the soil can significantly affect run-off; direct evaporation from the soil surface; the amount of soil moisture available to plants within range of their roots; and the depth to which roots can penetrate. How much water reaches each of these destinations over a given period depends on the physical condition of the soil and its influence on infiltration and run-off, and on the atmospheric conditions as they affect evaporation and transpiration.

CATCHMENTS AND WATERSHEDS

Water caught by a catchment will flow towards the lowest point at the outlet, where it may join water emerging from other catchments. The outer boundaries of a catchment are defined by ridgelines along the crests of the surrounding uplands. From the sides of a valley surface run-off tends to flow perpendicularly to the slope from crest to streamline.

A watershed is the area of land dividing two streamlines. Water moves away from the crest line towards the streamlines on either side. Thus a hill slope can be considered as either the inner slope of a catchment or the outer slope of the watershed. Catchments and watersheds are indicated on maps by the contour lines and by the course of drainage lines. Underlying geological formations, together with weathering and uplift processes, affect the form of landscapes. They influence the steepness or shallowness of slopes, whether the streamlines are of relatively sinuous shape or with abrupt changes of direction. The flow of water along the streamlines tends to cut the heads of streamlines back into the underlying materials.

For the purposes of enabling rainwater to soak into the soil and controlling the rate of flow of any excess run-off, we can subdivide a given catchment into a more detailed hierarchy of catchments, in which the smallest subdivisions may be measured in square centimetres, the larger in hectares, within catchments of square kilometres.

Rainfall entry into the soil depends on the porosity of the soil at any scale, while management of run-off and erosion across the surface also depends on any physical works that may be constructed when rainfall rates exceed even the best infiltration rate. Overlapping pairs of vertical aerial photographs viewed with a stereoscope provide a three-dimensional view of the landscape and surface features. A stereogram that shows a layout of roads along topographic crests and of conservation banks close to contours that has been designed in conformity with the natural catchments of the landscape.

This pattern provides a framework within which planting rows will have been contour-aligned relative to the conservation banks. The maintenance of soil porosity by mulch cover will allow the highest proportion of rainfall to be available as soil moisture for the crops and groundwater for the streams.

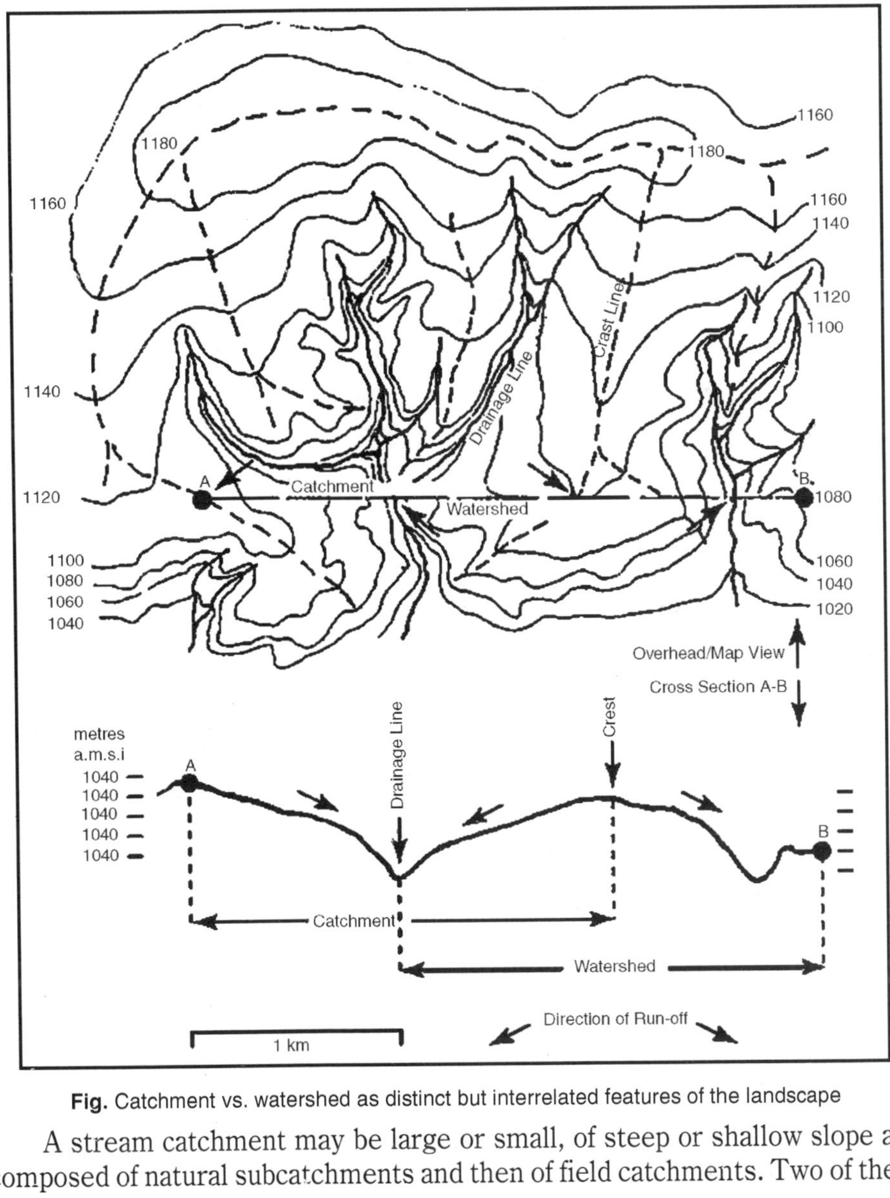

Fig. Catchment vs. watershed as distinct but interrelated features of the landscape

A stream catchment may be large or small, of steep or shallow slope and composed of natural subcatchments and then of field catchments. Two of these field catchments, which also form the left-hand side of a watershed whose crest runs along the ridge seen at top right.

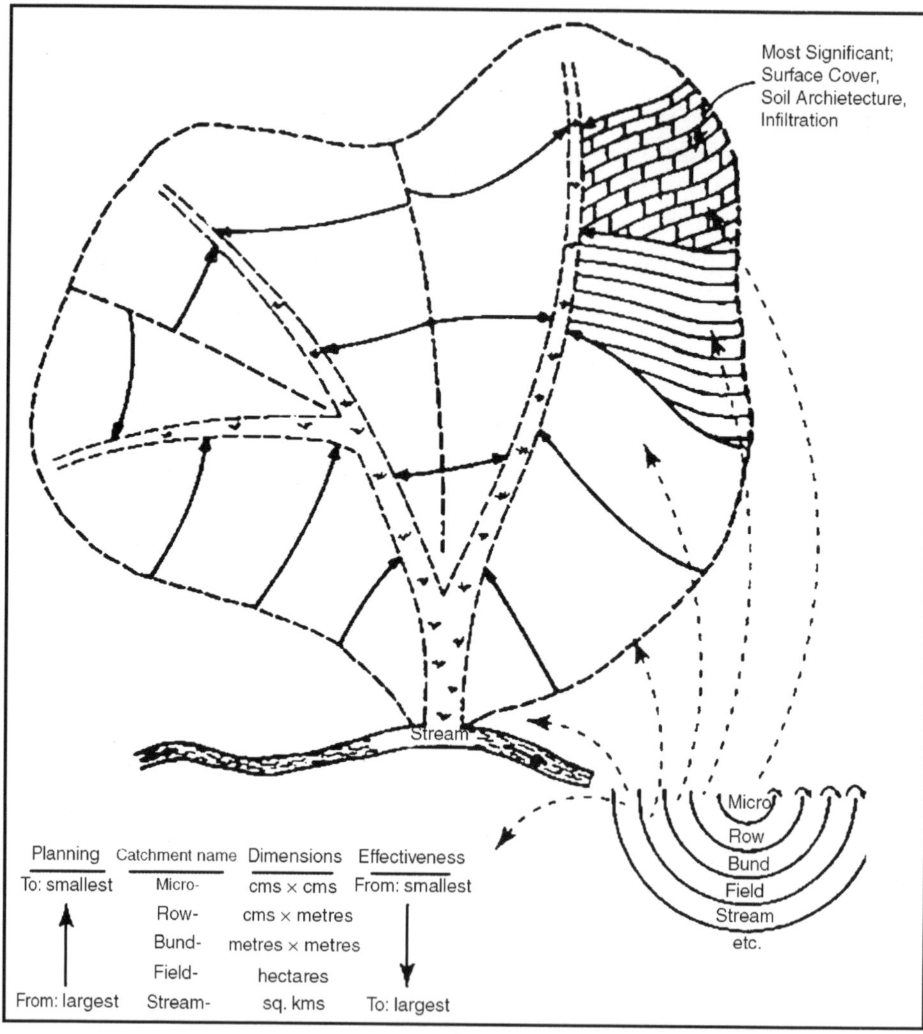

Fig. A nested hierarchy of interrelated constructed and natural catchments

In cropping agriculture the next smaller subdivision is the bund catchment, between any pair of physical conservation banks with its ridges and furrows (formal or informal) along planting rows. The physical conservation banks, vegetated with a fodder grass, subdivide the field catchment as well as separating the bund catchments. The aim of these structures is to conserve water and soil, though their effects on yield are disappointing. Their most significant function is to provide guidelines for contour planting of crop rows.

The smallest subdivision to trap rainwater and give it time to soak in is the microcatchment with its mulch.

In forestry, because the young trees are more widely spaced, the same effect can be obtained by a set of half moon shaped microcatchments, one at each planting position. Within this overall framework, the key to infiltration is to keep the soil porous with a cover of crop residues, which prevents damaging raindrop impact and provides a substrate for soil organisms.

The conservation effects of forests are due not so much to the presence of the trees themselves but to the litter of fallen leaves, twigs and branches, plus any low-growing vegetation. If the soil surface has not been damaged by trampling, less rainwater will run off and more will infiltrate into the soil.

How much plant-available soil moisture remains at a given time depends on the texture and porosity of the soil, the previous volume of soil moisture, the volume removed by direct evaporation, evapotranspiration and deep drainage. Irrigation (if available) is normally required when about two thirds of the available water - between field capacity (FC) and permanent wilting point (PWP) - has been depleted. If irrigation is not an option, it makes sense to manage the soil to develop and retain a maximum amount of soil pores of a wide range of sizes.

This will maximize the capacity for water retention and enable plants to withstand drought for longer periods. Loam textures generally have the largest available water capacity, while sand on the one extreme has a small available water capacity, as does clay at the other.

Available water capacity coupled with soil depth determines the volume of water usable by plants at a particular site. This is illustrated by comparing relevant characteristics and consequent amounts of available water for two soils on which tea is grown, one at Timbilil in Kenya, the other at Marikitanda in Tanzania.

Table. Differences in Available Water Capacities Between two East African Soils.

	Soil Depth (cm)	Bulk Density (g/cc)	Total Pore Space (% volume)	Depth of Water Held at FC (mm)	Depth of Water Held at PWP (mm)	Water Available to Plants (mm)
Timbilil (Kenya)	300	0.95	64	1 632	1 001	631
Marikitanda (Tanzania)	210	1.46	45	616	384	232

Flood flows in streams and rivers, which rise quickly after heavy rainfall derive mostly from rapid overland flow of water. Flood flows are often muddy with eroded materials. Clear streamflow originates from rainwater, which has infiltrated the soil and percolated through pores of a range of sizes at different slower speeds.

The streamflow during the rains from a cultivated catchment without any effective conservation measures and a nearby forested catchment, both on the slopes of the same mountain, within 1 km of each other.

The clear water, which runs throughout the year, more voluminous in the rainy than in the dry season, has percolated through the litter of fallen leaves and branches on the forest floor, which both protects the surface from rainfall impact and nourishes the soil organisms that maintain soil porosity. This water has travelled the slow route down through the soil to the groundwater, which moves into the stream via springs and seepages along the streambanks.

Fig. Mulching within a microcatchment.

Fig. The soil's water absorbing capacity in this *Eucalyptus* plantation is protected by the litter of leaves and twigs, and little run-off is likely to occur.

Fig. Much rainfall is likely to run-off this unprotected and probably compacted, bare surface beneath trees of the same genus *(Eucalyptus)* as in the previous photo; little rainwater will be able to infiltrate and soil moisture will be scarce.

Fig. Run-off and soil loss immediately after a rainstorm, Naisi catchment.

As dug wells provide direct access to shallow groundwater, on which many rural communities rely, it is important that enough rainwater penetrates and pass through the soil to replenish groundwater.

Exceptional rainfall such as during typhoons or hurricanes on already saturated soil can result in floods and erosion that change the landscape, no matter how well the land and the crops, grassland or forest are managed.

SOIL ARCHITECTURE AND THE IMPORTANCE OF PORE SPACES IN SOILS

Although we generally think of soil in terms of its solid parts, *i.e.* the sand, silt, clay and organic matter, it is the spaces between these solid particles that are as important as the solid particles. This is because it is the spaces where all the action takes place, just as in a house, where all the important activities occur in the rooms rather than in the walls and floors.

Fig. Dug wells such as this one may go dry because the water table falls below the depth to which the well was dug.

It is therefore the architecture of the soil that is important. The pore spaces in a soil vary in abundance according to the type of soil and how it has been managed. Soils under natural vegetation generally exhibit high porosity because of high biological activity and lack of interference by man. Consequently they have superior physical qualities compared with most soils used for crops or grazing. The contrasting porosity in forest and cultivated soils.

Pore spaces in soils vary in size, and both the size and continuity of pores have an important influence on the types of activities that occur in soil pores. The functions of pores of different size ranges, and their names, together with the size of crop roots.

Table. Functions and Sizes of Soil Pores.

Pores Size (mm Diameter)	Description of Pores	Functions of Pores
< 0.0002	Residual	Retain water that plants cannot use
0.0002 - 0.05	Storage	Retain water that plants can use *[PWP = 0.0002 mm; FC = 0.05 mm; but FC can vary from 0.03 to 0.1 mm diam. equivalent to 10 to 33 kPa]*
> 0.05	Transmission	Allow water to drain out and air to enter
> 0.1 to 0.3	Rooting	Allow crop roots to penetrate freely. *[Root sizes: seminal roots of cereals > 0.1 mm; tap-roots of non-cereals (dicots) > 0.3 mm; root hairs 0.005 to 0.01 mm]*
0.5 - 3.5	Worm holes	Allow water to drain out and air to enter
2 - 50	Ant nests and hannels	Allow water to drain out and air to enter

Pore sizes from 0.0002 to 0.05 mm diameter retain water that can be absorbed by crops and are referred to as storage pores, whereas smaller pores (the residual pores) hold water too tightly for plants to be able to extract it. Pores larger than about 0.05 mm diameter, referred to as transmission pores, allow water to drain through the soil and enable air to enter the pores as the water drains out.

Pore spaces are also needed for roots to freely penetrate soils in order to take up nutrients and water. The sizes of roots vary with the type of crop, but the smallest roots, apart from root hairs, have diameters of 0.1 to 0.3 mm and so soils must have pore spaces of at least this size if the smaller roots are to penetrate freely. In most soils roots grow partly through existing pores, the transmission pores, and partly by moving aside soil particles. Roots can only force their way into smaller pores if the soils are sufficiently compressible; the compressibility of soils increases with increasing water content, since water provides a form of lubrication between soil particles.

SOIL WATER MOVEMENT

The amount of water present in a soil, which is available for crop production, will depend on how much of the rainwater remains in the soil after the losses by run-off, evaporation, and deep drainage. The amount of rainfall that reaches the groundwater and thus contributes to water security, will depend on the extent to which the rainfall infiltrating the soil is in excess of that needed to replenish the soil's water holding capacity and satisfy the transpiration needs of the crops. Good rainwater management aims to maximize the amount of rainwater that enters the soil, and to make best use of it while it is there for use by crops and for recharging the groundwater. Any truly unavoidable surface run-off is conducted away safely in such a manner that it does not cause erosion problems.

When a well drained soil is saturated to the limit of its rooting zone, the rainwater that does not drain out of the root zone within 48 hours will be retained in soil pores smaller than about 0.05 mm diameter (the critical pore size may vary from 0.03 to 0.1 mm diameter). The quantity of water retained after 48 hours corresponds to the soil's field capacity (FC). The forces (or suctions) with which this water is held will vary according to pore size. The largest pores still to retain water will hold the water at about a tenth to a third of the pressure of the atmosphere (or 0.1 to 0.33 bar), depending on what suction corresponds to the soil's FC; this will vary with soil type and depth of the water table.

The maximum suction most crops can exert to withdraw water from soil varies with the crop, but the generally accepted value is equivalent to about 15 times the pressure of the atmosphere (*i.e.* 1.5 Mpa). This is approximately equivalent to the pressure that would be experienced when supporting a tonne weight on the palm of the hand. When soil water has been exhausted down to

15 bars, the water remaining in the soil will be that stored in pores smaller than 0.0002 mm diameter, and will correspond to the soil's PWP. Water held at suctions greater than the PWP is not available for plant growth. Consequently, it is water held between FC and PWP which can be used by crops for transpiration, and is termed the soil's Available Water Capacity (AWC). However, after a heavy rainstorm some of the water in excess of the soil's FC may be used by a crop while this excess water is percolating through the rooting zone.

The available water held within the range FC to PWP is retained with different strength, and about a third of it is not easily or rapidly available to crops, especially if the crops are transpiring strongly. The higher the transpiration demand, the more available (*i.e.* the less strongly held) the soil water must be to avoid crop water stress. In contrast, for a slowly transpiring crop even water held at higher suctions can be used without causing stress.

The maximum amount of available water that a soil can retain (*i.e.* the available water capacity) will vary with the soil's texture, organic matter content, rooting depth and structure. Soil organic matter is particularly important in that it can retain about 20 times its weight of water.

Organic soils and medium textured loamy soils with high contents of very fine sand and silt generally have the highest AWCs, clayey soils intermediate values, and soils with high contents of coarse sand the lowest AWCs. The stone content of soils can also be very important depending on the nature and abundance of the stones. Some ironstone gravel > 2 mm diameter can contain more than 20 per cent water (m^3/m^3) at FC and porous limestone and chalk can also make significant contributions to the AWC of a soil. In contrast, a high content of non-porous stones will greatly diminish the AWC of a soil.

For any given soil, the greater the rooting depth, the larger will be the quantity of soil water available to the crop. This is particularly important for annual crops as they have less time to develop deep and extensive rooting systems than perennial crops. The available water capacity may influence the length of growing period for crops grown on that soil. Soils of high available water capacity will permit longer growing periods because of their ability to provide greater quantities of stored water during dry periods than soils of low available water capacity. Shallow soils have little available water, and even in wet years they will be unable to benefit by storing any more water.

INFILTRATION OF RAINWATER INTO SOIL

In most areas where water shortages occur, maximizing the infiltration of rainfall into soil is indispensable to achieving food and water security. Land management should encourage infiltration as opposed to run-off. Exceptions are where rainwater harvesting is necessary for crop production and where high infiltration can lead to risks of landslides or other forms of mass movement.

The amount of rainfall that infiltrates will be governed by the intensity of the rainstorm in relation to the soil's infiltration rate. Excessive tillage and loss of soil organic mater often result in reduced infiltration rate due to loss of surface porosity. When storm intensity is greater than soil infiltration rate, run-off will occur, resulting in a waste of water that should have been used for crop production and for recharging the groundwater. The rate at which rainfall infiltrates into soil is influenced by the abundance, stability and size of the pores at the soil surface, their water content and by the continuity of the transmission pores into the rooting zone. In many soils the number of surface pores is rapidly reduced by the impact of raindrops, which break surface soil aggregates into small particles that clog surface pores and form surface seals or crusts with very few pores. The destructive raindrop action is avoided where there is a protective cover of crop foliage, residues, mulches or even weeds at or over the soil surface.

Other factors that can reduce the number, proportion and continuity of transmission pores are traffic by machinery, humans and animals, which destroys large pores by compaction, and tillage which disrupts the continuity of transmission pores through the smearing and compression of pores during plough pan formation in the subsoil. Infiltration rates are also affected by (1) the quantity of water present in the soil at the time of the rainstorm, which will depend on when the last rainstorm occurred and the permeability of the soil, and (2) the soil's capacity to retain water, which will vary with soil depth, stoniness, and texture.

PERCOLATION OF RAINWATER THROUGH SOIL

When a heavy rainstorm falls on a well-structured soil, rainwater percolates down through the dry soil as a wetting front, temporarily saturating the soil and displacing air. This is accompanied by the rapid drainage of water from the larger pores (greater than 0.05 mm) through gravity and the pressure of the mass of rainwater above. These larger pores exert only small forces of attraction on soil water. After about two days of drainage field capacity will have been attained and air will have re-entered the larger pores.

In poorly structured soils, rainwater will drain much more slowly. Drainage often continues for several weeks depending on the depth to the slowest horizon and the continuity of the larger pores with depth. In fine textured soils with cracks drainage water will flow down through the cracks in heavy rainstorms before the soil is saturated and while parts of the soil profile may still be dry. If the drainage water subsequently enters a smaller pore while passing through the soil, it will be retained, otherwise it will continue until it reaches the water table and contributes to the recharge of groundwater.

Once the drainage water has been lost from the rooting zone, further water movement within the root zone is slow and is referred to as capillary movement.

This movement is caused by forces of attraction, known as surface tension forces, which are exerted by soil particles on water. This movement can occur in any direction and includes the upward movement of water from water tables. Surface tension forces pull water into pores within the soil and the smaller the pores the more strongly the water is attracted and held.

Loss of Water Vapour from Soils

Water is also able to move through soils as water vapour. The most important example of this is the loss of water vapour by evaporation from soil surfaces. This occurs when the concentration of water vapour in the soil close to the surface is higher than that in the atmosphere immediately above. Water vapour will then move from the soil into the atmosphere. The drier and hotter the atmosphere compared with the surface soil, the greater will be the rate of evaporation from the soil, provided sufficient water can be supplied to the surface by capillary movement from below. Fine textured soils have an abundance of small pores and so more capillary movement of water to the surface will generally occur in fine textured than in coarse textured soils.

WATER MOVEMENTS INTO AND THROUGH A PLANT

Crops use large quantities of water, which under rainfed conditions come entirely from water in the soil, which in turn is derived from rainfall that has infiltrated the soil. A maize crop may use 400 to 750 mm of water depending on the rainfall and evaporation conditions. This corresponds to 4 000 to 7 500 cubic metres of water per hectare over the growing season.

Almost all the water absorbed from the soil by crop roots passes up through the stem into the leaves, where it evaporates and passes into the atmosphere in a process known as transpiration. This process accounts for almost all of the water absorbed by plant roots (about 99 per cent, the remaining 1 per cent being used directly in cell processes). Transpiration is essentially the same as the process of evaporation. Evaporation is what happens when a bowl of water is left in the sun. The liquid water disappears as it becomes converted into water vapour, and the higher the temperature, the drier the air, and the greater the wind speed, the greater will be the rate of evaporation. Evaporation occurs whenever water is exposed to the atmosphere, *i.e.* from lakes, rivers and puddles, and from the raindrops that accumulate on a leaf after a rainstorm. To ensure an efficient uptake of sufficient water by crops it is important that the crop roots are well distributed and able to penetrate deeply into the soil.

As a soil dries out from the surface downwards, so roots in the deeper layers tend to increase in number to compensate. When soil water reaches the surface of a root or root hair, it moves across the root into the xylem, which contains narrow tubes running through the root and extending up through the stems into the leaves. On reaching the leaves, water passes from the xylem into leaf cells where it evaporates into air spaces within the leaf. These air spaces are saturated with water vapour, and are connected to the normally drier outside air by very small openings in the leaves called stomata. During the day the stomata open, which allows carbon dioxide to enter the leaf. Sunlight is used to make sugars within the plant: a process known as photosynthesis. Part of the sugars are used to produce energy by a process called respiration, part are converted into substances forming the various plant organs.

Photosynthesis occurs only during daylight, whereas respiration occurs all the time. When the stomata open to allow carbon dioxide to enter, water vapour escapes into the drier air outside. For transpiration to occur there must be a continuous supply and movement of water from the soil to the plant to the atmosphere. The driving force responsible for this movement is the same as for evaporation, and can be simply described as the tendency for water to move, either as a liquid or a vapour, from where it is more abundant to where it is less abundant. In transpiration, water vapour moves from the very humid (*i.e.* high water vapour content) air spaces within the leaf into the drier atmosphere outside the leaf where the water vapour concentration is lower.

The movement of water vapour out of the leaf creates a suction (or "pull") on the water in the leaf cells, the xylem, the roots and the soil, so water moves into the root, up the xylem and into the leaves, to replace that which has been lost from the leaves. In addition to the transpirational suction, which causes water to move from the soil into the root, there is another force attracting water into the root known as osmosis.

In osmosis, water moves from where it is more pure to where it is less pure across a semi-permeable membrane. A semi-permeable membrane is a very thin skin, which has pores large enough for water to pass through into the root but not large enough for dissolved salts to pass out of the root.

Water therefore passes from the soil where the water is more pure (*i.e.* contains few dissolved salts) across the root surface (a semi-permeable membrane) into the root where the water is less pure (*i.e.* contains more dissolved salts).

Water Stress - Nutrient Interactions

Many areas with low and erratic rainfall where crop water stress is common are also deficient in nutrients, and the lack of nutrients is frequently the second most limiting soil factor. An interaction often occurs between soil water and nutrients, which means that soil water can influence the availability of nutrients, and the availability of nutrients can influence the uptake of soil water and a crop's resistance to drought. Thus, both factors can influence each other.

Plants contain a certain amount of water within them, which acts as a buffer against times of water shortage, but the amount is too small to last long. In contrast, plants store sufficient quantities of nutrients within their tissues to provide a buffer for longer periods when nutrients are not being absorbed. Consequently, water deficiencies become more quickly apparent and damaging than nutrient shortages. This suggests that conserving water may often be of prior and quicker benefit than attempting to conserve soil particles *per se.*

In addition, a lack of water also reduces the uptake of nutrients by a crop. This is largely because nutrients can only move to roots through water films within the soil, and so there must be continuous water films connecting the nutrients with the roots. A lack of soil water continuity, due to drought for example, will severely reduce the rate of nutrient uptake by crops. A lack of soil water will also diminish nutrient availability by reducing microbial activity, which is responsible for the liberation of nitrogen, phosphorus and sulphur from soil organic matter.

When there is a drought it is the surface soil, (which generally contains the bulk of the plants' roots and the soil's nutrients) which dries out first, and so while a crop may still be able to absorb water from the subsoil, it may suffer from a lack of nutrients.

A lack of available nutrients in the soil can restrict crop water uptake, especially when nutrients are limiting root development. This occurs most often in soils that are deficient in phosphorus. Applying P fertilizer to P-deficient soils will often promote root development, and as a result crop water uptake. Consequently the beneficial effects of applying P fertilizers are often relatively greater in seasons of lower rainfall than in those of higher rainfall.

The effects of drought and nutrient availability on crop yields are difficult to predict, because the effect will depend on when the water or nutrient shortage occurs in relation to the crop's stage of growth, and its needs and sensitivity to a lack of water or nutrients at that time. It is therefore often difficult to assess which factor, *e.g.* water or nutrients, is the more limiting to yield.

The most limiting factor can vary from season to season depending, for example, on when water shortages occur, and even during a season there will probably be periods when water is the main limiting factor, and other periods when nutrients are most limiting.

Water shortages often affect whether or not there is a response to fertilizers, and how much fertilizer should be applied. This is particularly common with N fertilizer, where the optimum response is frequently higher in good seasons than in poor seasons. For example, when there is no shortage of soil water an application of 40 kg/ha may prove to be the optimum application rate, but when water is lacking only 20 kg/ha may be the optimum amount to apply.

This creates difficulties in rainfed agriculture: since it is not possible to reliably predict the distribution and amount of rainfall, farmers cannot know how much fertilizer to apply. One approach that can help to overcome this problem is to apply a modest amount of N fertilizer at the beginning of the season assuming low rainfall, and then to apply additional quantities N later if the season appears promising.

Causes of Restricted Rooting

The most common cause of restricted rooting is physical restriction due to soil compaction, which results in the collapse or diminution of pore spaces and a localized increase in bulk density.

Once pores have been compacted to less than about 0.2-0.3 mm diameter, it is difficult for crop roots to freely penetrate the soil. Although the strength of compacted layers decreases as soil water content increases, a high water content can quickly limit the supply of oxygen to roots, so that roots then become restricted by a lack of oxygen.

Certain crops, such as cotton and sunflower, appear to be more susceptible to restricted rooting from compacted layers than others. Compaction often reduces pore sizes sufficiently to inhibit root penetration but not sufficiently to affect the drainage of water through the soil. Pores of 0.2-0.3 mm diameter can restrict roots but water can drain under gravity through pores as small as 0.01 mm diameter.

In mechanized cropping systems the continual use of tillage implements, especially disc ploughs, disc harrows, mould-board ploughs and rotovators, over long periods of time frequently results in the formation of dense plough pans containing few pores large enough to be penetrated by crop roots.

The plough pans develop just below the depth to which the soil is tilled and often have smooth upper surfaces with sealed pores, caused by the smearing action of mould-board ploughs. The degree of compaction depends on the pressure exerted by the implements on the soil.

Fig. The beginnings of a plough pan formed on a sandy clay loam after 8 years of disc ploughing and harrowing.

Land preparation when soils are wetter than the optimum moisture content for tillage promotes soil compaction, because the soils are then much more compressible. This is particularly likely to occur on soils that have deficient drainage, or are difficult to till in a dry state without pulverizing because of their very hard consistence (*e.g.* hardsetting soils). Compaction is also more likely when farmers use many passes to prepare the seed bed, or when they have only limited tractor power available and are unable to use wide sets of equipment and therefore produce compacted wheel ruts at closer spacing across the field surface. Compaction can also develop in the subsoil from the passage of heavy machinery such as combine harvesters and lorries loaded with grain, especially in wet conditions. The degree of compaction will depend on the total axle load of the machinery.

Soil compaction can also develop from hand tillage. Thin hoe pans just 2-3 cm thick can develop just below the depth of hoe penetration and thus restrict root penetration. When mounds or ridges are formed every year, the combination of hoeing at the same depth and the traffic of people within the furrows during wet conditions may accentuate the compaction. A similar effect to that of compaction can occur when structurally unstable soils, known as hardsetting soils, slump on becoming saturated by intense rainstorms to form dense layers. On drying the dense soil layer becomes very hard and restricts root penetration.

Restricted rooting may also be caused by naturally occurring dense horizons containing few pores large enough for roots to penetrate. These horizons may be found in soils formed from river, lake or volcanic sediments and in semiarid and arid areas where chemically cemented calcrete and gypsic horizons are formed.

In some situations root restriction can be caused by a seasonally fluctuating water table. During the rainy season the crop roots are confined to a shallow zone immediately above the high water table. If, when the water table falls during the dry season the crop roots have already completed their development, the roots will remain where they were, close to the surface, and without access to available water in the deeper subsoil. Roots may also be restricted to shallow depths by chemical factors, such as the presence of toxic aluminium or manganese, or by severe nutrient deficiencies in the subsoil.

INDICATORS OF RESTRICTED ROOTING

The most obvious indicator of restricted rooting when a crop is present is the distribution of the crop roots. When roots are physically restricted by a dense layer containing few pores suitable for root penetration, individual roots often develop characteristic growth patterns immediately above the restricting layer. The most common of these is the abrupt change in the direction of growth from vertical to horizontal, and a thickening of roots that do manage to penetrate the restricting layer just above the upper boundary of that layer.

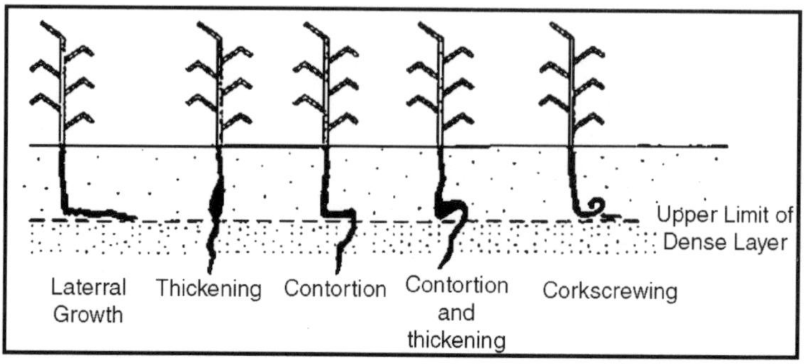

Fig. Examples of rooting pattern and growth habit when roots are physically impeded or prevented from penetrating a root-restricting soil layer.

In mechanized agriculture, plough pans are usually formed at 12-30 cm depth, depending on the implement used and its normal working depth. Naturally occurring dense layers may occur at any depth. The optimum time to observe roots is after flowering when most of the roots will have largely completed their growth.

When no crop is present, it is much more difficult to identify the existence of potentially root-restricting layers in a soil. However, the rooting pattern of mature weeds, either rooted or uprooted, that remain in the field after the crop has been harvested can be used to reveal the existence of a root-restricting layer. When neither a crop nor weeds are present, the presence of a dense soil layer of high strength and containing very few visible pores will often be a useful indicator. The presence of dense layers is often revealed when digging by the abrupt increase in resistance to the spade or hoe when the restricting layer is reached. However, sudden increases in soil resistance can also be experienced when the soil changes from moist to dry. To avoid this problem, it is advisable to wet the soil to 30 cm depth two days prior to carrying out the field examination.

Physically restricting layers can be identified by the scarcity of visible pores. The smallest pore visible to the naked eye (0.1 mm diameter) coincides reasonably well with the smallest pores into which the seminal roots of cereals (0.1 to 1 mm) and the tap-roots of dicotyledons (0.3 to 10 mm) can penetrate. When the density of visible pores observed in fragments of the dense layer from a soil pit is less than about six in an area of 10 cm × 10 cm, root restriction is likely to be severe, and responses to breaking up the restricting layer are likely.

Other indicators of potentially root-restricting layers that can be used in the field in the absence of a crop are soil strength determined with a penetrometer, and soil bulk density determined from undisturbed soil samples of a known volume. Critical penetrometer resistance and bulk density values at which the roots of most annual crops are restricted have been established for soils of different textures.

SOIL WATER

Water is the major transport agent for fluxes within and between terrestrial ecosystems. It is a prerequisite for all active life, and participates in geochemical cycles by weathering geological substrates, by leaching materials to groundwater and by moving ions and particles through the soil profile.

Soil water is part of the global hydrological cycle, which shows how water that falls as precipitation returns to the atmosphere through transpiration and evaporation.

On a local scale, how well the soil stores water is of great importance to crop production and the vitality of the land. Water and air constantly compete for pore space in soils. Most of the time soil pores are not full of water, but occasionally (after heavy storms or in swamps and bogs) some soils may be saturated with water and little or no air is present in pores.

SOIL WATER CONTENT

Soil water content on a volume basis (θ_v = volumetric water) is *the* ratio of water volume (V_w) to total soil volume (V_t):

$$\theta_v = V_w / V_t \text{ or}$$

$$\theta_v = V_w / (V_s + V_f)$$

V_s=volume of solids; V_f=volume of pores

Soil water content on a mass basis (θ_g = gravimetric water) is the ratio of mass of water (M_w) to total soil mass (M_t):

$$\theta_g = M_w / M_t$$

SOIL SOLUTION

Soil solution is the aqueous liquid phase of the soil and its solutes (dissociated ions).

Nutrients occur in the soil solution mostly as ions:

$$NH_4^+, NO_3^-, H_2PO_4^-, HPO_4^{2-}, K^+, Ca^{2+}, Mg^{2+}, SO_4^{2-}, Fe^{2+},$$

$$Mn^{2+}, Zn^{2+}, Cu^{2+}, MoO_4^{2-}, H_2BO_3^-, Cl^-.$$

Roots of higher plants can only take up nutrients from the soil solution. Uptake by plants and leaching by rain or irrigation depletes the amount of ions in solution unless they are replenished from reserves contained in primary minerals and organic matter and attached to colloids. Some processes of release are rapid (*e.g.* cation exchange), while others are slow with rates of release more critical (*e.g.* release of N, S and sometimes P by the decay of soil organic matter). The concentration of nutrients in solution and their rates of release depend upon the amount and types of solid-phase nutrients present in the soil, soil water content, temperature, aeration, microbial activity, and other properties of the soil solution.

SOIL WATER FLOW

The movement and retention of water, uptake and translocation in plants, and loss of water to the atmosphere are controlled by energy gradients. The forms of energy involved are potential (related to position), kinetic (related to movement), and electrical (related to cations and anions). The difference in energy level of water from one site or one condition (wet soil) to another (dry soil) determines the direction and rate of water movement in soils (and in plants).

Usually the energy status of soil water in a particular location in the profile is compared to that of pure water at a standard pressure and temperature, unaffected by the soil and located at some reference elevation.

The difference in energy levels between this pure water in the reference state and that of the soil water is called soil water potential (the term potential implies difference in energy status).

Soil water potential is due to several forces, each of which is a component of the total soil water potential (Ψ_t). Three main types of forces which contribute to the energy state of soil water include:

- Gravitational water has a positive energy and can flow out of the soil through the large pores.
- Osmotic potential is due to the attraction that salts have for water through the phenomenon of osmosis. This energy is negative relative to free water.
- Matric Potential, the potential energy of water attracted to soil solids. It too is negative.

In an unsaturated soil, matric potential results from the capillarity and adhesion forces. Plants must overcome the energy of matric potential to extract water from the soil.

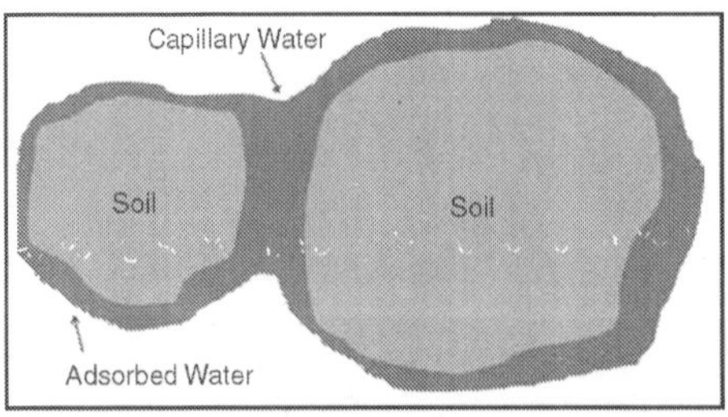

Fig. Matric Potential

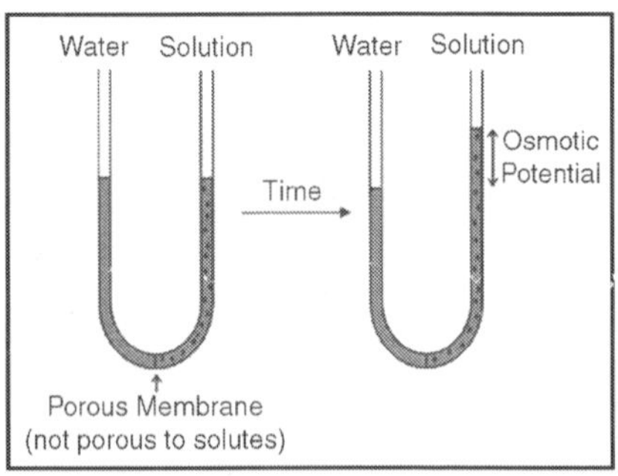

Fig. Osmotic Potentia

All of these components act simultaneously to influence water behaviour in soils. The general relationship of total soil water potential to potential energy levels can be expressed as:

$$\Psi_t = \Psi_g + \Psi_m + \Psi_s + \Psi_o + ...$$

Plants must overcome the energy of matric potential to extract water from the soil.

How do we know how much water is available in a soil?

For every soil there is a different distribution of pores of various sizes. Also, soils have different salt contents and particles. All of these factors will determine the energy at which the water is held in the soil and are reflected in the water retention curve.

The critical matric potential at which air enters the largest pores of a saturated soil is called the air entry value (AEV) for that soil.

The rate at which water flows is directly proportional to the water potential gradient, and the proportionality factor is called hydraulic conductivity (*K*). This relationship is described by Darcy's Law:

$$q = - K -(d_{\Psi}/dz)$$

where *q* is water flux density - the rate of water flow crossing a plane in unit time. The plane through which water flows is perpendicular to the axis of flow.

Note that water flux is the rate of water flow crossing the plane in m^3/s, while flux density is the flux divided by the area of the plane in $m^3/m^2/s$ or m/s.

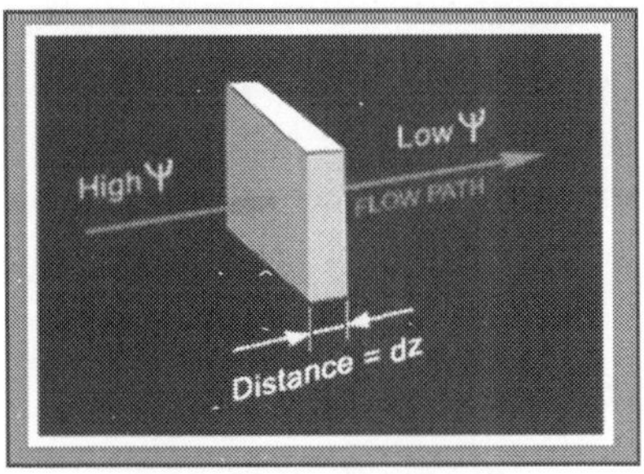

Fig. Darcy's Law

Illustration of water flow crossing the plane (note that the plane is always perpendicular to the axis of water flow".

Darcy's Law is a very powerful relationship, because it provides a basis for solving quantitative problems related to drainage, water supply to plant roots, etc.

Common principles of water movement in soils:

- Water flow is caused by a driving force resulting from a potential gradient ($d\Psi/dZ$)
- Water flow takes place in the direction of decreasing potential (hence minus sign)
- Rate of water flow (flux) is proportional to the potential gradient ($d\Psi/dZ$)
- Rate of water flow is affected by the geometrical properties of soil pores (size and tortuosity)

QUALITATIVE DESCRIPTION OF SOIL WETNESS

The terms used to qualitatively describe soil wetness refer to various stages along a continuum, and do not imply that soil water exists in different "forms". Despite the fact that these terms are qualitative and lack a precise scientific basis, they are widely used in practical soil management and help communicate important facts about soil water behaviour.

Maximum Retentive Capacity

When all soil pores are filled with water, the soil is saturated and it is at maximum retentive capacity. Matric potential is close to zero and the volumetric water content is essentially the same as total porosity (*f*). The soil will remain at maximum retentive capacity only as long as water continues to infiltrate, because water held in the largest pores (gravitational water) will percolate downward under the influence of gravitational forces.

Field Capacity

Once infiltration has ceased, water in the largest soil pores will drain downward quite rapidly. After one to three days, this rapid downward movement becomes negligible and matric forces begin to play a greater role in water movement. The soil is then at field capacity. In this condition, water has moved out of macropores and air has moved in. Micropores are still filled with water and continue to supply plants. Drainage continues but is much slower. Field capacity often corresponds to a matric potential of -1 to -3 m (or -10 to -30 kPa), assuming drainage into a less-moist zone of similar porosity.

Permanent Wilting Point (PWP)

Once an unvegetated soil has drained to field capacity, further drying is quite slow, especially if the soil surface is covered to reduce evaporation. However, if vegetation is present plants will continue to remove water. Water is first removed from the largest pores where water potential is relatively high. As these pores are emptied, roots will draw from progressively smaller pores in which matric water potential is lower and the forces attracting water to soil surfaces are greater. Therefore it will become more and more difficult for plants to remove water from the soil at a rate sufficient to meet their needs.

As the soil dries, the rate of plant water removal may fail to keep up with plant needs, and plants may begin to wilt during the daytime to conserve moisture. At first the plants will regain their turgor at night when water is not being lost through the leaves and the roots can catch up with plant's demand. Ultimately, the plant will remain wilted even at night when roots cannot generate water potentials low enough to coax the remaining water from the soil. Although not yet dead, plants below PWP cannot recover even if water is provided. For many herbaceous plants PWP corresponds to the water content retained by the soil when the matric potential is about -150 m (or -1.5 MPa or -1500 kPa).

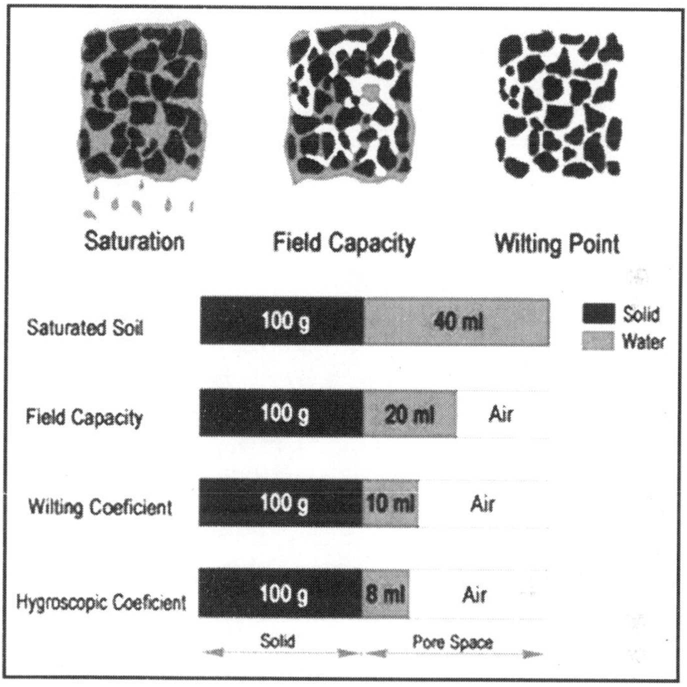

Fig. Water Content: Comparisons of a soil under saturated, field capacity, and wilting point, and hygroscopic point conditions

Available Water Storage Capacity (AWSC)

AWSC is considered to be the water retained in soils between field capacity and permanent wilting point (*i.e.*, between -10 to -30 and -1500 kPa):

AWSC = FC- PWP

Hygroscopic Water

Although plant roots do not generally dry the soil beyond permanent wilting point, if the soil is exposed to air, water will continue to be lost by evaporation. When soil water is lowered below PWP, the water molecules that remain are very tightly held, adsorbed ontocolloidal surfaces. This state occurs approximately when the atmosphere above a soil is essentially saturated with water vapour (98 per cent relative humidity) and equilibrium is established at a water potential of about -3100 kPa. The water is thought to be in films of 4-5 molecules thick and is held so rigidly that much of it is considered non-liquid and can move only in the vapour phase. Soil water content at this point is termed hygroscopic water. Soils high in colloidal materials will hold more water under these conditions than will coarse-textured soils.

2

Hydrological Operation in Agriculture Systems

DESIGN AND OPERATION OF AGRICULTURE

Water is one of the continuously renewable natural resources of the globe. The large cycle of the hydrosphere (the natural hydrological cycle) includes the evaporation of water from the surfaces of the oceans and the continents into the atmosphere, its return to the land-surface in the form of precipitation, and the surface and subsurface run-off conveying the water back to the lake basins, seas and oceans. It establishes a connection between the other spheres of the earth and it is an important component of the human environment. Economic development has created a second smaller cycle of water inside society. Water is taken out from natural resources, utilized in various forms and the effluents, having in most cases a modified quality, are released into the environment. The social cycle of water (water management) includes also those actions which modify the natural run-off to protect society against the harmful effects of water, including floods, erosion, and pollution.

Hydrology is a branch of water science describing the transport and storage of water through the cycles as well as the interactions between the water and its environment. Its task is to provide water management with information needed for designing and operating aquaculture structures and systems.

These systems modify the natural water regime according to the requirements of society, to avoid damages and to best utilize the water. Hence all users of water, aquaculture among them, need hydrological information produced by collecting and processing hydrological data as well as by analysing the processes described by these data.

Considering the task of hydrology, it is evident that not only the natural processes have to be analysed, but the modification of the water regime due to human intervention must also be investigated. Data from both the natural hydrological cycle and the use of water in the social sphere, therefore, should be collected and evaluated. The description of a water regime includes not only the determination of the quantities transported and stored, but also the qualitative properties of water. Hence the hydrological information systems must provide data about the instantaneous condition of and the expected changes in water quantity and quality. The information provided can be divided into two groups:

- Forecasts, prepared to inform the operators of existing hydraulic systems on the present condition of hydrological processes already in motion and the expected further development of these processes in a relatively short period (real time forecasting).
- Information required for planning, designing and constructing aquaculture structures and systems. This is composed of data on the extreme or average conditions of the hydrological processes expected to develop in the future without any limitation of time, other than the life span of the systems (design values).

In both cases forecasts have to be calculated from the hydrological data observed in the past. The basic difference between the two types of analysis is that in the first case the further development of a process started already is estimated and the actual time point of the occurrence of the predicted condition is determined, while the design values are calculated by using statistical methods and analysing long records without indicating the time of the development of the average or extreme conditions.

Hence hydrological investigations providing the basis of forecasting, and the determination of design values, together with the utilization of their results, should be discussed separately.

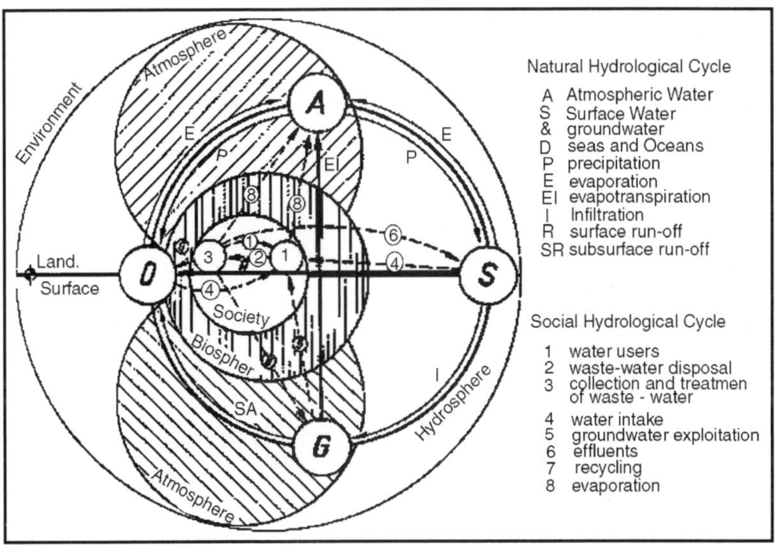

Fig. Natural and social hydrological cycles

There is another aspect which can be used to sub-divide the hydrological analysis, *i.e.*, the character of the processes. From this point of view the analysis, and the application of the information is basically different depending on where and how the data are gathered:

- Transport processes developing in the network of rivers with data measured at sections of the river, which are used for problems of water conveyance, sediment transport, and transport of dissolved solids.
- Areally distributed processes maintaining water exchange between the various water horizons and measured in the form of point values within the catchment such as precipitation, evapotranspiration, and infiltration.

Hydrological information needed for planning, design, construction and operation of any structures and systems of aquaculture, which follows:

Table. Hydrological Information needed for Operation and Design of Aquaculture Systems.

	Areally Distributed Processes (Vertical Water Exchange)	**Linear Transport Processes in Rivers (Horizontal Transport)**
Information for the operation of aquaculture systems (real time forecast)	quantitative forecast of precipitation	continuous forecasting of discharge and water level
	forecast of critical meteorological conditions (e.g., storm forecasting)	flood forecasting
		danger of accidental pollution
Determination of design data for planning and design of structures and systems	design parameters of climatic data	mean and variance of both the multi-annual averages and the expected extreme values of discharge and water level
	area-depth-distribution of precipitation	expected development of water quality and its relationship with water quantity and climatic data
	areal average of evapotranspiration	
	regime of soil moisture	
	balance of shallow ground-water	

HYDROLOGICAL FORECASTING FOR OPERATION PURPOSES

Forecasts of flow volumes and water elevation are essential to making the most efficient use of rivers and in minimizing damage due to floods. On rivers uncontrolled by dams and reservoirs, the river forecast is the basis of flood warnings, permitting removal of people, livestock and movable goods from the flood plain, and reinforcement of fixed river structures, such as bridges. On rivers controlled or partially controlled by dams built for power production, water supply, irrigation or flood control, reliable river forecasts permit operation of the dams for maximum benefit. River forecasts are essential to the reconciliation of joint use of reservoir capacity for water supply and flood control purposes. The flood control capacity in some cases must always be available to store flood run-off, or in other cases the reservoir must be drawn down in advance of a flood, to a level which will permit accommodation of the anticipated flood volume. When a flood occurs the reservoir is restored to its normal supply level. Accurate forecasts are obviously essential because, if the reservoir is lowered too much and the water supply level cannot be restored after the flood, the water users may be short of water. Contrarily, if not lowered enough, more water may have to be discharged than would have been necessary and flood damage may result.

All potential users of rivers can benefit from reliable river forecasts. Efficient operation of multi-purpose reservoirs and issuance of flood warnings, navigation and pollution control benefits can accrue. On some rivers bearing heavy pollution loads, regulation agencies have insisted on many industries and municipalities developing storage facilities to enable them to discharge wastes into rivers only when flows are great enough to give safe dilutions of the effluent. Forecasts of likely flows a day or two in advance permit scheduling of effluent discharge to minimize storage requirements and still keep pollution concentrations below a safe level.

Hydrological forecasting is, by definition, the prediction of the occurrence of a hydrological event specified both with respect to its quantitive measure and its actual time of occurrence. The following classification is made according to the purpose of a forecast:

	Quantity	Quality
Forecast of extreme situations	Flood, low-flow, ice, droughts	Sudden pollution
Continuous real-time forecast of possible future states	General purpose, for operation and control	

On headwater tributaries and drainage areas smaller than a few thousand square kilometres, precipitation observations and predictions must be used to produce river forecasts early enough to be of value for reservoir operation, flood warnings, etc. On the other hand, for larger rivers with longer lag times, forecasts can often be based to a large extent on observed upstream flows.

FORECASTING ON HEADWATERS AND SMALL RIVERS

For small drainage areas there are two main steps in the preparation of river forecasts. The first is to predict the volume of run-off by means of rainfall run-off correlations, and the second to forecast the distribution in time of the run-off volume. Rainfall run-off correlations can be derived from past records of storms and resulting river flows in a basin, either graphically or analytically. Such correlations permit forecasts of run-off volume from: (i) depth of storm rainfall over the watershed, (ii) a seasonal factor, (iii) an index of pre-storm moisture conditions in the basin, and sometimes (iv) storm duration.

Having determined the volume of rain and/or snowmelt which will run off, the timing of this run-off volume, and the peak flow that will result, must also be predicted. The unit hydrograph for the stream location in question is used to give the distribution of the run-off with time. Different methods are used in preparing the unit hydrograph, such as:

- Direct techniques (matrix method, use of orthogonal expansion, Fourier-series, etc.)
- Conceptual models (linear channels and reservoirs, Nash-cascade, soil-moisture accounting models, etc.)

For some purposes, such as reservoir operation, the unit hydrograph may be used to give a complete forecast of the rise and fall of a stream following a storm. For other purposes, such as flood warnings, and rapid forecasts to determine whether a given flood is likely to yield critical inflow rates to a reservoir, the important factors may be simply maximum stage and flow and the time at which they will occur.

It is possible to combine the rainfall-run-off correlations and the unit hydrograph peak values into a forecast graph which will give a rapid prediction of peak flow or stage. This is derived from the rainfall (or snowmelt) volume, the antecedent precipitation, a seasonal factor and the storm duration.

The trend in river forecasting is towards the use of physical parameters directly rather than through empirical correlations and index techniques. This has been made possible by the application of high-speed computers for rapid solution of more complex predictive equations.

FORECASTING THE REGIME OF LARGE RIVERS

Forecasts for large rivers and the lower reaches of principal tributaries generally make use of the dependable and consistent relationships between factors involved in stream-flow routing. The technique employed in a particular case depends upon whether the complete hydrograph is to be forecast or only the peak stage or discharge. On larger rivers the time between the end of the rainfall or snowmelt event and the occurrence of the peak of the resulting hydrograph at the point of interest is often measured in days, rather than hours as in headwater forecasting.

One of the simplest techniques in river forecasting makes use of the peak stage relation. This relation is easily constructed by plotting from past flood records the peak stages at the upstream station against the corresponding peak stages at the downstream station. The success of such a relation in providing an accurate estimate of the peak water level at a downstream station is dependent upon two assumptions: (i) the volume of inflow between the upstream and downstream station is small in comparison with the flow at the upstream station or is a constant proportion of that flow; and (ii) the peak of the local inflow always bears the same time relation to the peak of the flow at the upstream station.

A complementary relation is the time-of-travel curve which, as the name implies, provides an estimate of the time for the peak stage to travel from the upstream to the downstream station. The time of travel may vary directly with the stage throughout the range of stage or it may begin to increase more rapidly or less rapidly after a certain stage is reached depending upon the stream channel and valley characteristics.

Headwater hydrographs derived from rainfall-run-off correlations and unit hydrographs can be translated to downstream points on the main river by the techniques of streamflow routing. These techniques are essentially simplifications of the gradually varying unsteady flow equations.

DETERMINING DESIGN PARAMETERS

An important application of hydrometeorology is the provision of criteria for the design of various water control structures, such as dams and storage reservoirs, storm sewers, bridges and irrigation systems. In order to determine the design values statistical techniques are used.

Hydrology and meteorology are concerned with natural phenomena, and the analysis of problems arising from these phenomena are dependent upon observations of their magnitude, duration and location. There will be little likelihood of obtaining successive measurements of the same event. Using statistical methods, a multiplicity of data can be reduced to manageable form and so enable the relevant data to be distinguished from the irrelevant. Statistical methods are particularly useful in checking the validity of a hypothesis involving complex natural phenomena.

Having formulated an hypothesis, a statistical analysis of whether the observations fit that hypothesis would show whether or not there are good reasons to doubt it. Statistical analysis deals with probabilities and cannot provide conclusive proofs.

The procedures for estimating design values can be grouped as follows:

- Frequency analysis of flood flow records or frequency analysis of precipitation records and relating of precipitation to run-off from the catchment using the unit hydrograph or some other method.
- Transposition of storms.
- Transposition of depth-area-duration relationships.
- Probable maximum precipitation (PMP) method.
- Regional methods.
- Empirical methods.

Determination of both probable maximum storm and standard project storm depends on the idea of storm transposition. The purpose of storm transposition is to increase the storm experience of a basin by considering not only storms which occurred in that basin, but also those which released their heaviest rainfall on adjacent areas that are meteorologically 'similar'. There are two difficult problems in storm transposition.

The first is to define the region of meteorological similarity over which a particular storm can be transposed. The second concerns the permissible change in orientation of the storm rainfall pattern to yield critical run-off rates and volumes.

RESERVOIR DESIGN

One of the design problems which is common to most uses of water is the degree to which the seasonal and annual fluctuations of streamflow can be modified by storage to correspond with the demand requirements; that is, the problem of gearing the supply to the demand by means of storage reservoirs. A reservoir design requires solutions to many problems.

The storage capacity of a reservoir may be required for one or several of the following uses: irrigation, municipal and industrial water supplies, power, flood control, sediment storage, navigation, recreation, and fish and wild-life habitat. However, in a given drainage basin the amount of storage available to serve one or a combination of these uses may be limited by economic considerations or by physical limitations such as topography, foundation strength for supporting a dam, and water supply. Economic factors involve such items as incremental costs of additional capacity, cost of right-of-way, cost of transportation facilities and many other factors peculiar to a particular reservoir location. Field surveys will provide data on reservoir capacity and reservoir area related to elevation of the water surface at the dam site. These non-hydrologic data are necessary for relating available supply to probable demands for the water.

The first hydrologic problem is to determine the supply of water that will be available for the proposed project. The best information on supply would be obtained from long-term flow records at the reservoir site. Records from a station a short distance upstream or downstream from the point in question would also be valuable. If adequate data are not available, the existing record can be extended or streamflow data synthesized.

An important consideration in determining reservoir capacity is the minimum annual run-off. The available storage determines the magnitude of demand that can be met during a period of low run-off. It is desirable, therefore, that the record or estimate of stream-flow volume embraces at least one dry period, which may encompass from one month to several years. Most reservoirs operate on an annual cycle, with each year's run-off, modified by storage, capable of meeting the demand. Some reservoirs are operated so that storage is carried over for several years to be certain of meeting a specific minimum demand even during a most severe drought.

In addition to deficient precipitation, drought may be associated with increased evapotranspiration losses.

During such periods the flow in many drainage basins is from groundwater so that the flow in the stream is closely related to groundwater depletion. If the groundwater should drop below the bed of the stream, then the flow of the stream will cease except for surface run-off from snowmelt or intense local precipitation. At high latitudes, minimum flows commonly occur during the months associated with long periods of freeze-up. The precipitation is held in storage in the form of ice and snow and the moisture in the soil layers may be frozen. In these regions temperatures are well below freezing over large areas, therefore run-off conditions also exhibit a fair degree of uniformity. This enables more than usually reliable comparisons to be made between adjacent streams and hence estimates of low flows at ungauged sites may be made more readily.

WIND EFFECTS

On lakes and reservoirs, both wind and differences in atmospheric pressure can cause changes in water levels which may seriously affect human safety, design and operation of dams, dikes, breakwaters, water intakes, and other lake structures, and the flow of outlet rivers.

The wind exerts a horizontal stress on the water, raising the level from normal at the leeward shore and lowering the level on the windward shore. This is called a set-up effect which lasts as long as the force continues to be applied by the wind and is sometimes known as a wind tide. However, if the wind dies down quickly, or changes direction, an oscillatory motion takes place with alternate high and low levels being observed at both ends of the lake, with a periodicity characteristic of the dimensions of the water body. The oscillations are known as seiches.

The calculation of the magnitude of the initial set-up may be an important factor in assessing the amount of freeboard allowance required on dams, dikes, embankments or other retaining structures. It should be noted that set-up is greatest on long, shallow lakes and least on small, deep water bodies.

The transport by lake currents of water and materials in suspension and solution may have practical implications. Pollution from municipalities along lake shores and from ships on a lake will move with the currents. Under unfavourable conditions such pollution may enter water intakes. Currents also cause erosion of the shoreline. For these and other reasons, it is often necessary to know something of the currents in a lake or reservoir and the factors which cause them. The theory of currents in an infinite water body, applicable to the oceans, is also useful in dealing with lakes.

Wind-induced currents in a lake are closely interlinked with seiche and set-up effects because there must be currents to permit motion of large masses of water. When the wind which produces a set-up dies down, a current will flow from leeward to windward to restore the lake to its natural level.

However, the current will continue due to its inertial energy and result in the temporary piling up of water at what was the windward end while set-up winds prevailed. This produces the seiche oscillations previously discussed. These oscillations are thus accompanied by currents, sometimes very strong ones.

Littoral drift is the transport, primarily by waves and coastal currents, of beach material along the shores of large lakes and reservoirs. As waves approach a coastline in a direction determined by the direction of the wind which generated them, they are refracted, since the portion of the wave first reaching shallow water slows down first. This tends to change the wave direction to one more parallel to the bottom contours and thus to the shoreline. However, this is rarely completely achieved and the breakers still retain a component of velocity along the shore, and shore currents are established. These currents can move eroded material along the shore.

DETERMINATION OF AREAL PRECIPITATION

Precipitation is the input to the continental branch of the hydrological cycle. Its best known forms are rain and snow, although there are areas where the amount of water reaching the surface in other forms (*e.g.* dew, fog, etc.) is not negligible either.

The amount of precipitation is measured by various standardized rain gauges, which might be supplemented with recording and telemetering devices. Without going into the detailed analysis of the reliability of gauge-readings, it is necessary to mention that there are systematic errors (due to evaporation, splashing, aerodynamic effects, wetting the instrument), apart from the random ones caused by erroneous readings or by the defects of instruments. The observed amount of precipitation is, as a rule, generally smaller than the true value.

In spite of the systematic errors of rain gauges the determination of the depth and intensity of precipitation at the observation points (point values) is more reliable than the measurements of other hydrological processes (*e.g.*, river discharge or evaporation). More serious uncertainties are involved when the areal average of the precipitation is calculated, than those disturbing the basic data.

The information required for any hydrological study dealing with either surface run-off or groundwater regime is the total amount of precipitation reaching the ground surface of the catchment within the time interval investigated. The observation of precipitation provides, however, only point-values (the sample caught by one gauge having an orifice of 200 cm).

The most important task of the hydrological analysis of precipitation data is, therefore, to calculate areal averages from gauge readings by constructing area-depth-distribution and to consider the change of precipitation between the measuring level and the ground surface.

Considering the problems explained, the actual input of continental hydrological systems should be improved by ensuring a double transformation of the measured precipitation data.

The most efficient increase of the accuracy of hydrological computation can be achieved by:

- Calculating the average of the areal precipitation from point values (investigating also the structural character of the field of precipitation); and
- Reducing the measured values by considering the losses from precipitation between the measuring level and the terrain (characterization of interception).

THE REGIME OF SOIL MOISTURE

The soil moisture zone is the heart of the continental branch of the hydrological cycle, because its structure and instantaneous condition determine the ratio of surface to subsurface run-off and even a considerable amount of evapotranspiration is also covered from the water stored below the terrain and above the water table. The lower boundary of the soil moisture zone is the water table, which is a theoretical surface, where the total pressure prevailing in the water phase of the porous medium is equal to the atmospheric pressure. The excess pressure, which is the difference between total and atmospheric pressures, is, therefore, equal to zero. Hence the basic physical difference between soil moisture and groundwater is that the former is under suction because the pressure prevailing in the water above the water table is smaller than the atmospheric pressure.

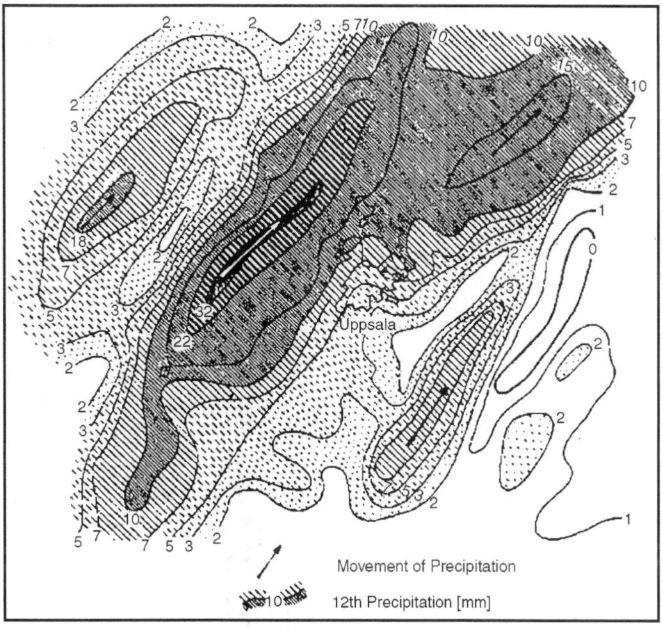

Fig. Area - depth - distribution

It is necessary to consider that the water content of a porous medium is a function of the suction of the soil moisture. Hence both the storage capacity (soil moisture retention curve) and the transport capacity (unsaturated hydraulic conductivity) must be investigated depending on the suction distribution in the soil moisture zone. The field measurements should provide, therefore, information not only on the vertical distribution of moisture content, but the suction value must be measured at several points along the vertical profile. From this double observation both the change of the stored amount of water and the direction of the propagation of the moisture can be determined.

In practice the information required most frequently for solving engineering problems includes the infiltration through the surface and the water exchange between soil moisture and groundwater (groundwater accretion). The former is needed to calculate surface run-off, which is the difference between precipitation reaching the terrain and infiltration. Ground-water accretion provides information on the vertical recharge and drainage of groundwater systems. Therefore both infiltration and groundwater accretion require the simulation of the complete soil moisture system.

The detailed analysis of the regime of soil moisture has two main parts:

- The theoretical investigation should give methods to determine the interrelationships between suction and water content (soil moisture) retention curve or pF curve) as well as between suction and unsaturated hydraulic conductivity. Substituting these basic relationships into the transport equation (Richards equation) and combining the condition of continuity and the resistivity against the movement in unsaturated porous media in this way, the regime of the soil-moisture zone can be simulated. The model is suitable to determine any practical information required for hydrological studies such as infiltration, groundwater accretion, and instantaneous condition in and transport through the soil moisture zone.
- The practical part of the study deals with the observations and the measuring methods applied to determine the hydrological processes developing between the land-surface and the water table. The most important components are the vertical distribution of both the moisture content and tension, the amount of infiltration through the surface and the water exchange between soil moisture and groundwater.

EVAPORATION AND EVAPOTRANSPIRATION

Evapotranspiration is the main form of drainage from the continental branch of the hydrological cycle. Among its two components evaporation is basically a physical process of transforming a liquid into vapour. When this transformation is made by plants (transpiration) the process is influenced also by biological, biophysical and biochemical factors.

Unfortunately there is no direct method to measure the vapour flux, and indirect observations provide only point values similar to the measurement of precipitation. For the determination of the areal average of evapotranspiration the whole system transforming liquid water into vapour should be investigated.

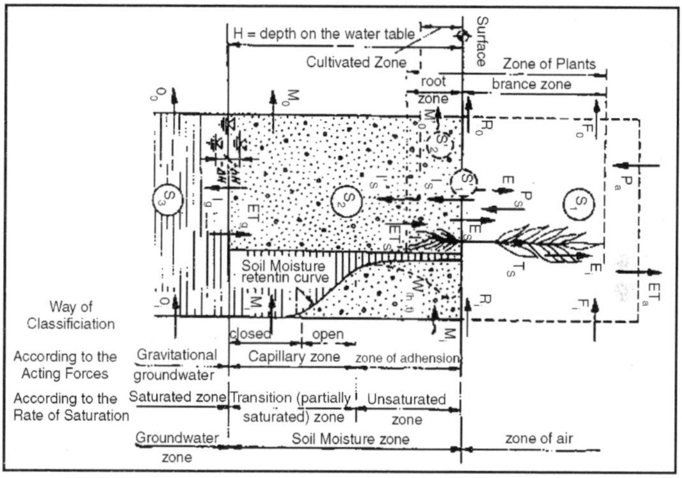

Fig. Soil moisture system*Hence the investigation of evaporation and evapotranspiration includes the following:*

It is divided into two subsystems: (i) the vapour receiving air-mass; and (ii) the vapour producing terrain. The character of the interface between the two subsystems where the evapotranspiration actually occurs, and the amount of energy available here, are important factors in the system. The availability of water and the rate of saturation of the air may be also limiting factors.

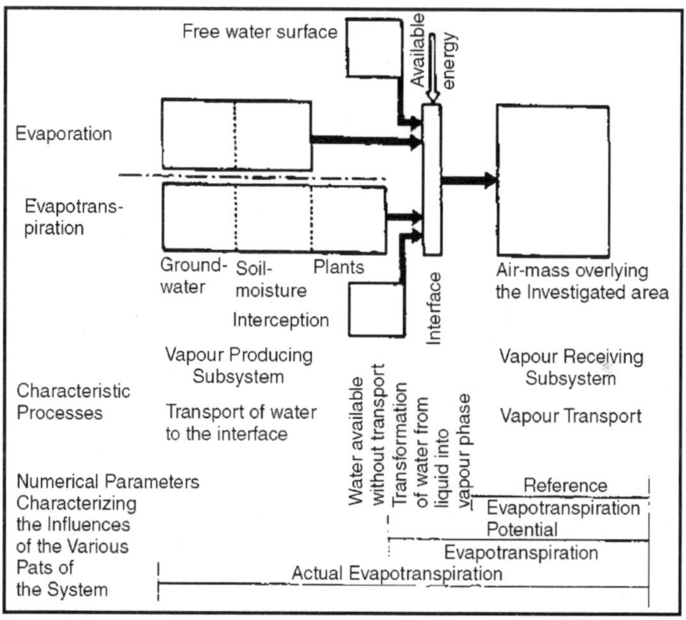

Fig. Evapotranspiration system

To design and operate hydraulic structures and systems the areal average of actual evapotranspiration should be determined. For this calculation the amount of water actually drained from each part of the catchment having different surface-conditions should be estimated.

- Measurement of evaporation from free water surface by use of evaporation pans. The determination of the pan-to-lake coefficient considering the oasis effect due to energy advection must be taken into consideration.
- The main components of evapotranspiration. To determine the best methods suitable for measuring evapotranspiration, distinction should be made according to the origin of water evaporated and transpired, *i.e.*, interception, soil moisture, groundwater.
- Methods to estimate actual evapotranspiration. The use of lysimeters, the measurement of the decrease of stored water and the calculation of evapotranspiration from the deficit of heat balance are investigated.

GROUNDWATER PROBLEMS ARISING IN CONNECTION WITH AQUACULTURE STRUCTURES

Groundwater, when it is under natural conditions, should be in balance, *i.e.*, the multi-annual averages of its recharge and drainage respectively have to be equal. If not balanced, continuous rising or lowering of the water table would be observable. Any human intervention disturbing this equilibrium will have far reaching consequences. The exploitation of groundwater introduces a new drainage activity without changing the natural inputs and, therefore, it causes the continuing depletion of the groundwater resource. Aquaculture structures which raise the level of surface waters (*e.g.*, dams or river barrages, canals, fish ponds) change the natural conditions, initiating artificial recharge or decreasing the natural drainage maintained by percolation from groundwater reservoirs to rivers. This change causes a rise in the water table until the development of a new equilibrium.

The design of aquaculture structures requires a forecast of expected changes in the groundwater regime due to the structures, followed by continuous observation after they are built. When the aquaculture structure modifies the movement of groundwater only in its close vicinity, the application of the principles of seepage hydraulics gives sufficient results, while more complicated groundwater systems have to be simulated to determine the far reaching effects of large engineering structures. In this case the hydrological processes interconnecting the groundwater systems with the surface and atmospheric branches of the hydrological cycle should be considered as boundary conditions. For the investigation of complicated groundwater systems, it is necessary to include:

- The geological structure of aquifers and aquitards using geological mapping, geophysical prospecting and drilling.
- Hydraulic conductivity and storage coefficient, calculated from the physical parameters of the layers and measured either in laboratory or in the field.
- The observation of potential-head by using test wells.
- The boundary conditions influencing the development of seepage in the system.

EFFECTS OF AGRICULTURAL PRACTICES

HYDROLOGY

Knowledge of how water moves through a catchment is fundamental to understanding the effects of agricultural practices.

One of the primary objectives of this work, therefore, was to develop a conceptual model of the sources, distribution, and movement of water within the DR2 catchment. A simple water balance, estimated from 2003 data, provides insight into the overall hydrology of the catchment and the processes that influence agricultural effects on the environment in this study area.

Irrigation water, imported from the Yakima River, is the principal source of water to the catchment, accounting for nearly 60 per cent of the total inflow. This irrigation water is delivered through the Sunnyside Canal and is transported throughout the catchment by a series of secondary canals, from which it is applied to crops and pasture. Rill irrigation—the predominant application method used in the catchment—typically results in 15 to 25 per cent of the applied water running out the end of the furrows (U.S. Environmental Protection Agency, 2003; Eisenhauer and others, 2006; Schwankl and others, 2007). This tailwater is routed into surface or subsurface drains that eventually discharge to regional surface-water drains, such as DR2. Sprinkler irrigation—the other major water application method in DR2—produces no tailwater in most circumstances. Most canal water diverted into the irrigation water delivery system is applied to the land surface; however, a small fraction of the diverted water goes unused as part of the normal operation of the gravity-driven irrigation-water delivery network. This end-of-system "operational spill" also is routed into the regional drainage network. Additional unused water may enter the drainage system due to a lag time between requests for changes in the delivery amount and the implementation of the request, the onset of unexpected cool or wet weather, system malfunctions (such as pipe breaks), and various on-farm exigencies.

Other primary sources of water to the catchment are ground water that flows into the area from north of the Sunnyside Canal, precipitation, and leakage from the Sunnyside Canal. Together, these four major sources contribute approximately 180 cm of water per year to the catchment. Evapotranspiration and stream discharge account for nearly 90 per cent of the total catchment outflow. The third important component of outflow is ground water.

Irrigation is a key component of the hydrology of the DR2 catchment. In fact, there is no evidence that a surface-water system existed in the catchment before irrigated agriculture began in the area more than a century ago. In addition to creating an extensive surface-water drainage network during the growing season, long-term irrigation has caused ground-water levels to rise, and the higher water table provides base flow that sustains the surface-water flow year-round in the southern part of the catchment.

A hydrograph of the discharge of DR2 near the catchment outflow shows the influence of the annual irrigation cycle. Ground-water levels measured in wells throughout the catchment were consistently higher than the stage of the water in DR2, indicating that ground water flowed towards DR2 at all times. During the non-irrigation season (mid-October through mid-March), base flow in the stream is sustained by ground water, most of which originated as infiltrating irrigation within and upgradient of the DR2 basin. During the irrigation season (mid-March through mid-October), discharge at site W13 increased two- to threefold. The shape of the annual discharge hydrograph is defined by the controlled application of irrigation water, and because of the absence of major precipitation events, short-term variation in stream flow is small and occurs predominantly during the irrigation season, driven by crop demands and agricultural practices.

A conceptual model of subsurface flow in the DR2 catchment was developed on the basis of several lines of evidence, including ground-water levels, ground-water age estimates, the isotopic composition of water from selected sites, visual observations of flow in the surface drainage network, and results of numerical ground-water modeling and flow-path simulations. In this conceptualization, regional ground-water generally flows from north to south, but the local shallow system is more complex, consisting of small-scale subsystems driven by individual irrigation events and constrained by local topography and the extensive system of artificial drains (both surface and buried) that intercept shallow ground water and route it towards the surface-water drainage network. While the drain system promotes lateral flow to the stream, upward flux of deeper ground water through the streambed is limited, especially in the southern part of the catchment, by low-permeability streambed material. At the streambed transects, permeabilities ranging from 10^{-5} to 10^{-6} m/s were estimated by temperature flux modeling. Finally, a considerable volume of deep ground water flows out of the catchment as ground water, not discharging to the surface system until it reaches Granger Drain or the Yakima River.

Water-level measurements in wells in the northern part of the catchment—at sites W24 and W25—consistently showed a downward gradient. Several lines of evidence indicate that a downward vertical gradient also exists in the groundwater system near well W30. Although a second well at a different depth was not available to confirm it, the seasonal fluctuations in the ground-water level and temperature at W30 were similar to those in W24 and W25. In addition, age tracers (SF_6 and CFCs) in water from W30 were consistent with concentrations in the atmosphere in the early 1980s, indicating little mixing of this water with either older or younger water. Finally, there are no surface water discharge zones in the area. All these factors are consistent with predominately downward vertical movement of ground water.

Ground-water levels in the northern part of the catchment fluctuated on an annual cycle, with the lowest levels occurring in late spring to early summer, and the highest levels occurring in late summer to early fall in a delayed response to irrigation. The annual range in water levels in these wells was 1–2 m. This range is larger than was observed downgradient because the relatively thick unsaturated zone in the upper catchment prevents discharge to surface-water outlets. Although the water table rises and falls in the upper portion of the catchment in response to infiltrating irrigation water, the response to individual irrigation events is dampened during the movement of the water through the thick unsaturated zone. Water temperatures in these wells also fluctuated on an annual cycle, but the seasonal variation was less than 1°C, again due to the insulating and moderating influence of a relatively thick unsaturated zone.

In the southern part of the catchment, at sites W20 and W21, as well as at neighbouring site W32, ground-water levels showed an upward gradient. At these wells, water levels sometimes fluctuated on a daily basis, but showed no seasonal pattern. The daily fluctuations were likely a response to individual irrigation events and (or) evapotranspiration, evident in these wells because the water table in this part of the catchment is relatively shallow. The shallow water table also was reflected in the water temperatures in these wells. Relative to those in the upper catchment, temperatures varied considerably, with summer maximum temperatures approximately 5°C greater than winter minimum temperatures.

The midcatchment, where the predominant direction of vertical flow in the ground-water system transitions from downward to upward. Comparing levels measured in sites W23p and W23q shows that the vertical gradient is generally upward, but periodically reverses to downward during the irrigation season. Data from this well demonstrate the response of the ground-water system to individual irrigation events and illustrate changes in the flow system in response to a change in irrigation method. Rill irrigation was used near these wells during 2003. In 2004, the farmland immediately upgradient of these wells was converted to sprinkler irrigation. During 2003, the frequency of head reversals was greater and the magnitudes of downward gradients were greater. Water levels in these wells also illustrate the general seasonal effect of irrigation—the vertical gradient diminished when canal deliveries ceased in mid-October 2003, and approached zero by the end of the end of the winter. Finally, corresponding increases in water temperature indicate that gradient reversals in the ground water at site W23 are not due simply to a pressure response, but rather to warmer water from the surface or near surface actually being transported into the saturated zone and raising the local water table.

Estimated dates of ground-water recharge, based on interpretation of the CFC, SF_6, and ^3H data, generally are consistent with the conceptual model of subsurface flow developed from ground-water levels. Data from wells in the midcatchment suggest that this is a mixing zone of waters recharged at different times. Age tracers from site W24q are consistent with this water being a mixture of water recharged around 2000 and water recharged around 1970.

Age-tracer data from site W23q are consistent with water in this well being predominantly water recharged around 1980 but indicate contributions from water recharged prior to the mid-1960s. These mixtures are consistent with vertical mixing that would result from periodic vertical gradient reversals in the midcatchment and also with the conceptual model of a dynamic shallow flow system highly influenced by irrigation applications and drainage modification sitting atop a reservoir of considerably older, slower moving water. Throughout the catchment, water from the shallow water-table wells contained at least a component of recently recharged water, reflecting the movement of excess irrigation water into the underlying saturated zone.

The effects of artificially enhanced drainage in the southern part of this system can be seen by comparing the hydrographs of wells at sites W20p and W21p. Short-term fluctuations in the water level at W20p show the effects of individual irrigation events in the adjacent pasture. By contrast, the water level at W21p shows little variation, likely due to the damping effect of the shallow drain that surrounds this field and efficiently shunts excess irrigation to the surface-water system. Age-tracer data from W21p, which indicate that the shallow ground water here was not recently recharged, support this reasoning.

The stable isotopes of water from selected sites also corroborate the conceptual model of flow in DR2. From late April to late August, the canal water becomes enriched in the heavier isotopes, oxygen-18 (^{18}O) and deuterium (^2H). Samples from W23p and W24p are isotopically indistinguishable from late season canal water, consistent with this shallow ground water being recently applied irrigation. W21s is somewhat more depleted in the heavy isotopes than the other wells sampled. This was the deepest and oldest ground water sampled, and this isotopic signature likely reflects a contribution from older, regional ground water. W25p is slightly enriched in the heavier isotopes relative to other samples; these isotope values suggest a source of water similar to other nearby ground water (such as that from W23p, W24p, and W25q) that has undergone additional evaporative losses.

The isotopic composition of streambed water from W63c differs considerably from all other water sampled. That sample is much more depleted in ^{2}H and ^{18}O than other water samples collected for this study, indicating that the shallow subsurface water near the catchment outlet is derived from a different source than other water sampled. Particle-tracking simulations suggest that the deepest, longest ground-water flow paths in the catchment discharge near W63c and that water from the streambed there includes a significant component of the oldest ground water in the catchment. Tritium data for the three streambed sites sampled (W51c, W55b, and W64c) also indicate that water from these sites includes a substantial component of relatively old water, recharged more than 50 years ago.

The isotopic signature of water from the DR2 drain (W13) is similar to that of shallow ground water and water from the Sunnyside Canal, and the shift in isotopic signature from spring to summer in water from the drain mimics that seen in water from the canal. During both seasons, however, the water from DR2 drain was slightly depleted in the heavier isotopes relative to corresponding canal water samples, which is consistent with a small contribution from deep ground water.

WATER CHEMISTRY

Water chemistry data were collected at sites representing several environmental compartments. Analysis of samples from the Sunnyside Canal (W15) indicates the composition of imported irrigation water prior to distribution in the catchment. Samples from a field drain (W42) provide an example of the chemistry of excess irrigation run-off from a single field—in this case likely a mixture of overland flow and shallow soil drainage. Samples from wells (W20–W32) reveal the characteristics of ground water at various locations throughout the catchment, both near the water table and somewhat deeper, but all in the shallow part of the flow system.

Sites in two streambed transects in the lower reach of DR2 (W51–W65) show the composition of ground water just prior to its discharge into the stream. Finally, samples from near the DR2 catchment outlet (W13) reveal the composition of surface-water discharge from the catchment.

As discussed earlier, virtually all of the surface-water discharge from the catchment can be attributed to excess irrigation. As this water passes through the catchment, its chemistry is altered by biochemical and geochemical reactions at the land surface, in the soil, and in the ground-water system.

Fertilizers and other soil amendments used in the DR2 catchment are important sources of N and P, as well as potassium, calcium, magnesium, sodium, chloride, and sulfur. Cattle manure is applied as a fertilizer and soil amendment to many crops in the area either as liquid slurry or as solids, and, based on the classification scheme of Kendall (1998), ^{15}N and ^{18}O isotopes of nitrate suggest that manure may be a major source of the nitrate measured in waters throughout the catchment.

Mineralogical analyses of subsurface material (sampled during well drilling) show its composition is predominantly quartz, plagioclase, potassium feldspar, and calcite. Through weathering processes, these minerals are abundant sources of silica, calcium, sodium, and potassium. Dolomite and hornblende, which contain magnesium, also are common in the area. Finally, smectite, which has a high cation-exchange capacity, was an important component of the colloidal fraction of most samples analyzed, and likely plays a role in water chemistry.

Comparing the chemistry of water from a number of environmental compartments and locations provides insight into the processes that affect the quality of water discharging from this catchment:

- *Irrigation water*: From the Sunnyside Canal had the lowest concentrations of major solutes and dissolved nutrients of any compartment sampled. Because the canal is the major source of water to the catchment, the chemistry of water from other environmental compartments gives an indication of how this canal water is altered as it passes through the catchment.
- *Field drain*: The field drain flowed year-round, and both the discharge and chemical character of the water fluctuated with time. During winter, the major solute chemistry of water from the drain was similar to that of water from nearby well W21p, suggesting that winter flow in the drain was discharging ground water. Over the course of the irrigation season, the chemistry of the water from the field drain varied considerably (for example, N ranges from less than 1 to more than 23 mg/L; silica ranges from 13 to 47 mg/L), suggesting changing contributions from several sources, including ground water, shallow soil water, and irrigation tailwater. This composite discharge from the drain consistently had higher concentrations of major solutes and dissolved nutrients than water from the canal. Concentrations of the major solutes calcium, magnesium, potassium, sodium, chloride, sulfate, and silica, were inversely related to drain flow rates (R^2 = 0.4–0.6). Concentrations of nitrate also tended to be lower at higher flow rates (R^2 = 0.3), but although the highest concentrations of soluble reactive P (SRP) were measured during low-flow conditions, there was no apparent relationship between SRP and flow ($R^2 < 0.1$). The major solutes, as well as nitrate, also were inversely correlated with concentrations of suspended sediment in water from the field drain, though relationships were weaker than for discharge (R^2 = 0.2–0.3). The highest concentrations of SRP also occurred when concentrations of suspended sediment were low, but again, the relationship between SRP and suspended sediment was not significant

($R^2 < 0.1$). Data from Ebbert and others (2003) and Fuhrer and others (2004) point to a relatively constant contribution of SRP from ground water in this area. The poor correlation between SRP and suspended sediment may reflect differing mixtures of shallow ground water, soil water, and overland flow, in which concentrations of suspended sediment and dissolved constituents can vary considerably, depending on the amount of time since the field was last tilled and the timing of sample collection relative to individual irrigation events. Longer contact times between soil and water at lower flow rates may also be a contributing factor to the patterns observed. Durand and others (1999) noted that increased depth and velocity of overland flow (such as would occur during times of high drain discharge) tended to decrease concentrations of solutes.

- *Ground water*: The chemistry of the water from most individual wells varied little over the course of the study, but differences in water chemistry among wells were substantial. Relative to water from the canal and field drain, ground water typically had higher concentrations of major solutes. Although concentrations varied considerably among wells, water from the shallow wells typically had higher concentrations than that from deeper wells, suggesting that water-rock interactions in the subsurface may be a less important source of the major solutes than near-surface sources such as cattle manure, other fertilizers and soil amendments, and weathered soil.

The highest concentrations of N in ground water, which was present predominantly as nitrate, generally were detected in water from wells in the upper catchment. Concentrations of N in water samples collected from well W25p consistently exceeded 70 mg/L.

This is probably the result of a complex interaction of factors, including the local ground-water flow system and local agricultural practices. Site W25 is at a local topographic high, and subsurface flow is likely to be directed downward rather than laterally.

The very recent recharge date calculated for W25p is consistent with relatively rapid, direct downward flow near the water table. Furthermore, this well is in an area where cattle manure, a potential source of considerable N, was applied liberally.

Concentrations of excess dissolved N_2 in water from most wells was low. Dissolved N_2 in excess of the concentration achieved through equilibration with the atmosphere plus excess air indicates that nitrate previously present in the water has undergone denitrification—the process by which nitrate is transformed into N_2.

Thus, low concentrations of excess dissolved N_2 in water from most wells indicate that denitrification is not a major process in this ground-water system.

Green and others (2008) present the details of excess dissolved N_2calculations and discuss the limited denitrification in this ground-water system.

Streambed transects.—Similar to the findings for ground water, the chemistry of water from individual sites within the streambed transects varied little over the course of the study. However, the range in the data from each transect shows that water chemistry differs over very small distances (tens of centimeters) in this system. Some of these differences may result from spatial variations in the composition of the streambed, but the variability observed also suggests that flow paths from different source areas are converging to discharge into this lower reach of DR2, and provides insight into differences between the chemistry of relatively deep ground water that discharges upward into the stream and that of the shallow ground-water system that flows laterally into the stream.

Compared to the ground water sampled, water from both streambed transects tended to have lower specific conductance and alkalinity, as well as lower concentrations of calcium, magnesium, and sodium. In contrast, concentrations of potassium, chloride, silica, and SRP tended to be similar to those measured in ground water. These conditions indicate that differences between water in the streambed and ground water were not likely due to simple geochemical processing or dilution (from unapplied irrigation water, for example), but suggests that a component of the streambed water comes from a source distinct from other ground water sampled.

We hypothesize that this source is relatively old ground water discharging from the deep, long ground-water flow paths mentioned previously.

The concentrations of nitrate were lower in water from the streambed than in most ground water sampled. In water samples collected from the upstream transect, concentrations of excess N_2 consistently exceeded 2.5 mg/L, except at sites W62b and W61b. In water samples collected from the downstream transect, concentrations of excess N_2 ranged from less than 1 to more than 6 mg/L (at site W54a). These data indicate that denitrification occurs in portions of the subsurface, but that conditions vary over very small distances. Values of ^{18}O and ^{15}N isotopes from nitrate collected from the streambed also suggest that denitrification occurs in this zone.

Although several lines of evidence indicate that denitrification occurs in this zone, the effects on nitrate concentrations in surface water are limited by the small flux of water involved.

Several factors combine to constrain water flux through the streambed. First, as mentioned earlier, the permeability of the streambed is low; second, irrigation flow during the growing season raises the water level in DR2, which reduces the gradient from ground water to the stream; and third, enhanced shallow drainage throughout the catchment shunts ground water laterally towards surface drains and effectively bypasses deeper flow paths.

Conditions at sites W61b and W62b, in the shallow, western part of the upstream transect indicate that lateral flow to the stream from a distinct water source occurs in this area. Concentrations of nitrate here were higher than in the rest of the transect, but concentrations of excess N_2 were somewhat lower, suggesting a source of nitrate-rich water, but only limited progression of denitrification (due to either reduced capacity or shorter residence time).

These two sites also had higher specific conductance, alkalinity, calcium, magnesium, potassium, sodium, and total N (mostly present as nitrate), and lower concentrations of chloride relative to other sites in the transect. Furthermore, water temperature near these sites was periodically colder than water deeper in the streambed (sites W61c and W62c) or water near the bottom of the overlying stream. These data indicate that colder water was flowing laterally towards the transect at middepth and provide evidence of lateral ground-water flow to the stream.

- *Catchment outlet:* The chemistry of the water at the catchment outlet is consistent with the water being a mixture of canal water, field drainage, ground water, and streambed discharge. Examining the chemistry at the catchment outlet in the context of the annual streamflow hydrograph provides further insight into processes in the catchment. Several of the major solutes (calcium, magnesium, sodium, chloride, sulfate, and silica) were inversely related to stream discharge. This pattern of lower concentrations during the irrigation season, when the highest discharges occurred in DR2, is illustrated by calcium and silica data. Concentrations of nitrate in water from DR2 also were highest during the non-irrigation season, but concentrations during the irrigation season varied more than those of the major solutes. In contrast, concentrations of SRP in the catchment outflow were highly variable, but generally increased during the irrigation season.

3

Fundamentals of Remote Sensing

PROCESS OF REMOTE SENSING

The process of remote sensing involves the detection and measurement of radiation of different wavelengths reflected or emitted from distant objects or materials, by which they may be identified and categorized by class/type, substance, and spatial distribution. The background required for use of remote sensing tools may seem overwhelming at first. The decisive factor in the successful application of remote sensing data, however, need not be the technical sophistication of the user, but rather the suitability and precise use of the tool to obtain accurate and relevant data. A general grasp of the technical process that transforms electromagnetic energy into useful information can improve and expand the appropriate use of these tools. Nevertheless, depending on the application, social scientists wishing to work with remote sensing imagery would do well to partner with physical scientists with a deeper understanding of how the imagery represents physical processes on the ground.

In the broadest sense, remote sensing refers to information gathered by measuring and interpreting signals. To use perhaps the simplest analogy, the human body is constantly involved in a variety of remote sensing tasks. Hearing and vision are two obvious examples, involving the gathering and interpretation of sound and light waves, both in limited ranges of the entire sound and light spectra.

Active and *passive* remote sensing are used to describe the way sensors gather data. To illustrate the two types of sensors, consider a snapshot camera, a sensor that captures electromagnetic radiation in the visible spectrum. Outdoors in full daylight, a camera is ordinarily used as a passive sensor in that it receives reflected visible light from its surroundings and uses optics, a shutter, and film to create a lasting image. At night, on the other hand, when there is inadequate light for most cameras to capture a useable image, the camera may employ a flash. The flash is emitted from the camera and bounces off the object to illuminate it, just as an active (or radar) sensor sends a burst of energy towards its target and then receives the reflected radiation. A key factor in the choice between passive and active sensor is the relative strength of the potential signal each system must measure. For a passive system, the source of the signal is ultimately the sun, which emits electromagnetic radiation at its highest intensity between the ultraviolet and infrared ranges. In the radar wavelength ranges, however, sensors must provide a signal of sufficient intensity to travel to the earth, and return with enough strength to be distinguishable from the background "noise" from other sources.

Energy Source or Illumination (A)-A basic requirement for remote sensing is an energy source to illuminate or provides electromagnetic energy to the target of interest. For passive instruments, this is usually the sun; for so-called active instruments, the sensor itself emits an pulse of energy.

Radiation and the Atmosphere (B)-As the energy travels from its source to the target, it will come in contact with and interact with the atmosphere it passes through. This interaction also takes place a second time as the energy travels from the target to the sensor. Inevitably there is a certain degree of atmospheric scattering of radiation.

Interaction with the Target (C)-Once the energy makes its way to the target through the atmosphere, it interacts with the target depending on the properties of both the target and the radiation.

Recording of Energy by the Sensor (D)-After the energy has been scattered by, or emitted from the target, a sensor (remote-not in contact with the target) collects and records the electromagnetic radiation.

Transmission, Reception, and Processing (E)-The energy recorded by the sensor has to be transmitted, often in electronic form, to a receiving and processing station where the data are processed into an image (hardcopy or digital).

Interpretation and Analysis (F)-The processed image is interpreted, visually or digitally/electronically, to extract information about the target illuminated.

Application (G)-The final element of the remote sensing process is achieved by applying the information that has been extracted from the imagery about the target in order to better understand it, reveal some new information, or assist in solving a particular problem.

FUNDAMENTAL PROPERTIES AND CONCEPTS

THE ELECTROMAGNETIC SPECTRUM

All objects including plants and soil emit and or reflect energy in the form of electromagnetic radiation. Electromagnetic radiation travels in waves propagating through space similar. Three major components of these waves are frequency, amplitude and wavelength. Frequency is the number of cycle crests passing a point during a given period of time. One cycle per second is referred to as one hertz. Amplitude is the energy level of each wave measuring the height of each wave peak. Wavelength is the distance from the top of one wave peak to the top of the following wave peak The most common source of electromagnetic radiation that we are familiar with is the sun. The sun radiates energy covering the entire electromagnetic frequency spectrum.

Remote sensors act similar to the human eye. They are sensitive to images and patterns of reflected light.

A major difference between the human eye and remote sensors is the frequency range of the electromagnetic spectrum that they are sensitive to. The electromagnetic spectrum range varies from very short wavelengths of less than ten trillionths of a meter known as gamma rays, to radio waves with very long wavelengths of several hundred meters.

The electromagnetic spectrum can be sliced up into discrete segments of wavelength ranges called bands, also sometimes referred to as a channel. It is the sun that most often provides the energy to illuminate objects.

The sun's radiant energy strikes an object on the ground and some of this energy that is not scattered or absorbed is then reflected back to the remote sensor. A portion of the sun's energy is absorbed by objects on the earth's surface and is then emitted back into the atmosphere as thermal energy.

Visible Region

The visible light portion of the electromagnetic spectrum ranges from 0.4 micrometers ("μm") (shorter wavelength, higher frequency) to 0.7 μm (longer wavelength, lower frequency). This is the frequency range of light that the human eye is sensitive to.

Every object reflects, absorbs and transmits electromagnetic energy in the visible portion of the electromagnetic spectrum and also other non-visible frequencies. Electromagnetic energy which completely passes through an object is referred to as transmittance.

Our eyes receive the visible light reflected from an object. The three primary colours reflected from an object known as *additive primaries* are the blue, green and red wavelengths.

Primary colours cannot be formed by the combination of any other primary colours. Intermediate colours are formed when a combination of primary colours are reflected from an object.

Magenta is a combination of reflected red and blue, cyan a combination of reflected blue and green, and yellow a combination of reflected red and green. Colour film produces colours by using layers of dyes which filter out various colours. The three colours which absorb the primary colours, known as *subtractive primaries*, are magenta, cyan and yellow. Magenta absorbs green and reflects red an blue, cyan absorbs red and reflects blue and green, and yellow absorbs blue and reflects red and green. The absorption of all colours produces black. If no colour is absorbed then the film produces white.

Infrared Region

The non-visible infrared spectral region lies between the visible light and the microwave portion of the electromagnetic spectrum. The infrared region covers a wavelength range from.7 μm to 14 μm. This broad range of infrared wavelengths is further subdivided into two smaller infrared regions. Each of these regions exhibits very different characteristics. The infrared region closest to visible light contains two smaller bands labeled near infrared and short-wave infrared with wavelengths ranging from.7 μm to 1.1 μm, and from 1.1 μm to 3.0 μm respectively. These infrared regions exhibit many of the same optical characteristics as visible light. The sun is the primary source of infrared radiation, which is reflected from an object. Cameras used to capture images in the visible light spectrum can capture images in the near infrared region by using special infrared film.

The other infrared region with longer wavelengths ranging from 3.0 μm to 14.0 μm is composed of two smaller bands labeled mid-wave infrared and long-wave infrared with wavelengths ranging from 3.0 μm to 5.0 μm, and from 5.0 μm to 14.0 μm respectively. Objects generate and emit thermal infrared radiation thus these objects can be detected at night because they are not dependent on reflected infrared radiation from the sun. Remote sensors operating in this infrared wavelength range measure an object's temperature.

AN OVERVIEW OF REMOTE SENSING

Remote sensing is the examination or the gathering of information about a place from a distance. Such examination can occur with devices (*e.g.*-cameras) based on the ground, and/or sensors or cameras based on ships, aircraft, satellites, or other spacecraft. Today, the data obtained is usually stored and manipulated using computers. The most common software used in remote sensing is ERDAS Imagine, ESRI, MapInfo, and ERMapper.

Modern remote sensing began in 1858 when Gaspard-Felix Tournachon first took aerial photographs of Paris from a hot air balloon. Remote sensing continued to grow from there; one of the first planned uses of remote sensing occurred during the U.S. Civil War when messenger pigeons, kites, and unmanned balloons were flown over enemy territory with cameras attached to them. The first governmental-organized air photography missions were developed for military surveillance during World Wars I and II but reached a climax during the Cold War.

Today, small remote sensors or cameras are used by law enforcement and the military in both manned and unmanned platforms to gain information about an area. Today's remote sensing imaging also includes infra-red, conventional air photos, and Doppler radar. In addition to these tools, satellites were developed during the late 20th century and are still used today to gain information on a global scale and even information about other planets in the solar system. For example, the Magellan probe is a satellite that has used remote sensing technologies to create topographic maps of Venus.

TYPES OF REMOTE SENSING DATA

The types of remote sensing data vary but each plays a significant role in the ability to analyze an area from some distance away. The first way to gather remote sensing data is through radar. Its most important uses are for air traffic control and the detection of storms or other potential disasters. In addition, Doppler radar is a common type of radar used in detecting meteorological data but is also used by law enforcement to monitor traffic and driving speeds. Other types of radar are also used to create digital models of elevation.

Another type of remote sensing data comes from lasers. These are often used in conjunction with radar altimeters on satellites to measure things like wind speeds and their direction and the direction of ocean currents.

These altimeters are also useful in seafloor mapping in that they are capable of measuring bulges of water caused by gravity and the varied seafloor topography. These varied ocean heights can then be measured and analyzed to create seafloor maps. Also common in remote sensing is LIDAR-Light Detection and Ranging. This is most famously used for weapons ranging but can also be used to measure chemicals in the atmosphere and heights of objects on the ground.

Other types of remote sensing data include stereographic pairs created from multiple air photos (often used to view features in 3-D and/or make topographic maps), radiometers and photometers which collect emitted radiation common in infra-red photos, and air photo data obtained by earth-viewing satellites such as those found in the Landsat programme.

APPLICATIONS OF REMOTE SENSING

As with its varied types of data, the specific applications of remote sensing are diverse as well. However, remote sensing is mainly conducted for image processing and interpretation. Image processing allows things like air photos and satellite images to be manipulated so they fit various project uses and/or to create maps.

By using image interpretation in remote sensing an area can be studied without being physically present there. The processing and interpretation of remote sensing images also has specific uses within various fields of study. In geology, for instance, remote sensing can be applied to analyze and map large, remote areas

Remote sensing interpretation also makes it easy for geologists in this case to identify an area's rock types, geomorphology, and changes from natural events such as a flood or landslide. Remote sensing is also helpful in studying vegetation types. Interpretation of remote sensing images allows physical and biogeographers, ecologists, those studying agriculture, and foresters to easily detect what vegetation is present in certain areas, its growth potential, and sometimes what conditions are conducive to its being there.

Additionally, those studying urban and other land use applications are also concerned with remote sensing because it allows them to easily pick out which land uses are present in an area. This can then be used as data in city planning applications and the study of species habitat, for example. Finally, remote sensing plays a significant role in GIS.

Its images are used as the input data for the raster-based digital elevation models (abbreviated as DEMs)-a common type of data used in GIS.

The air photos taken during remote sensing applications are also used during GIS digitizing to create polygons, which are later put into shapefiles to create maps.

Because of its varied applications and ability to allow users to collect, interpret, and manipulate data over large often not easily accessible and sometimes dangerous areas, remote sensing has become a useful tool for all geographers, regardless of their concentration.

PROCESS OF GATHERING INFORMATION IN REMOTE SENSING

Remote sensing refers to the process of gathering information about an object, at a distance, without touching the object itself. The most common remote sensing method that comes to most people's minds is the photographic image of an object taken with a camera. Remote sensing has evolved into much more than looking at objects with our eyes. It now includes using instruments, which can measure attributes about objects which unaided human eyes can't see or sense.

Some other definitions of Remote Sensing are: "Photogrammetry and Remote Sensing are the art, science and technology of obtaining reliable information about physical objects and the environment, through a process of recording, measuring and interpreting imagery and digital representations of energy patterns derived from non-contact sensor systems".

"Remote sensing may be broadly defined as the collection of information about an object without being in physical contact with the object. Aircraft and satellites are the common platforms from which remote sensing observations are made. The term remote sensing is restricted to methods that employ electromagnetic energy as the means of detecting and measuring target characteristics".

"Remote sensing is the art and science of obtaining information from a distance, *i.e.* obtaining information about objects or phenomena without being in physical contact with them. The science of remote sensing provides the instruments and theory to understand how objects and phenomena can be detected. The art of remote sensing is in the development and use analysis techniques to generate useful information".

HISTORY

In 1858 a French photographer, Gaspaed Felix Tournachon was the first to take aerial photos from a tethered balloon. A few years later in 1861, aerial photographs became a tool for military intelligence during the civil war.

Aerial photographs were also taken from cameras mounted in kites (1858), and on carrier pigeons (1903). In 1909 Wilber Wright flew the first airplane to take the first photographs in flight.

The first aerial photographs used in the process of creating maps was presented in a paper in 1913, by Captain Tardivo at a meeting of the International Society for Photogrammetry. Military aerial photos were used on a large scale during World War I. The military trained hundreds of people to process and interpret aerial reconnaissance photos.

The French aerial units developed 56,000 photos in four days during the Meuse-Argonne offensive in 1918. After World War I and through the 1930's, commercial aerial survey companies employed many former military personnel to process aerial photos to produce maps such as topographic maps, forest management maps, and soil maps.

World War II saw the development of colour-infrared film for the US Army in 1942. These images were used to detect enemy forces and equipment that were camouflaged. A majority of Allied intelligence gathered about the enemy during this war was the direct result of aerial photoreconnaissance.

The U.S. military and other government agencies such as National Aeronautics and Space Administration (NASA) continued to develop the use of remote sensing during the cold war years. The 1960's also saw the expansion and development of earth remote sensing from space.

The first military space photo reconnaissance satellite, Corona, was launched in 1960. Corona took pictures of the Soviet Union and its allies using photographic film. The exposed film was then transferred into unmanned recovery vehicles in space.

The recovery vehicles then de-orbited and returned to earth by parachute carrying the film, which was then processed and analyzed in the lab. The first series of weather satellites called the Television Infrared Observation Satellites (TIROS) began launching in 1960. NASA continued collecting images for its earth observation surveys, from outer space, with the Apollo and Gemini spacecraft.

Scores of U.S. meteorological and earth observation satellites were launched during the 1970's. Also during the 1970's manned spacecraft such as the Skylab space station collected images of earth from outer space. In 1972 Landsat-1, with an original resolution of only 80 meters was the first satellite launched into space for non-military earth resource observation. Landsat contained sensors capable of taking multispectral digital images. U.S. military photoreconnaissance satellites have been keep secrete and unavailable to the general public. Starting in 1976 the U.S. military started deploying more sophisticated high-resolution satellites capable of relaying digital images to earth. Eight Keyhole-11 satellites were launched between 1976 and 1988. Three improved Keyhole-11B satellites were launched between 1992 and 1996. They are able to produce images with estimated resolutions of nearly ten centimeters (four inches).

Non-military satellite images have been used to monitor the degradation and pollution of the environment. These images also can be used to assess the damage of floods and natural disasters, assist in forecasting the weather, locate minerals and oil reserves, locate fish stocks, monitor ocean currents, assist in land use mapping and planning, produce geologic maps, and monitor range, forestry and agricultural resources.

PRINCIPLES OF REMOTE SENSING

We perceive the surrounding world through our five senses. Some senses (touch and taste) require contact of our sensing organs with the objects. However, we acquire much information about our surrounding through the senses of sight and hearing which do not require close contact between the sensing organs and the external objects. In another word, we are performing Remote Sensing all the time.

Generally, Remote sensing refers to the activities of recording/observing/ perceiving (sensing) objects or events at far away (remote) places. In remote sensing, the sensors are not in direct contact with the objects or events being observed.

The information needs a physical carrier to travel from the objects/events to the sensors through an intervening medium. The electromagnetic radiation is normally used as an information carrier in remote sensing. The output of a remote sensing system is usually an image representing the scene being observed. A further step of image analysis and interpretation is required in order to extract useful information from the image. The human visual system is an example of a remote sensing system in this general sense.

In a more restricted sense, remote sensing usually refers to the technology of acquiring information about the earth's surface (land and ocean) and atmosphere using sensors onboard airborne (aircraft, balloons) or spaceborne (satellites, space shuttles) platforms.

SATELLITE REMOTE SENSING

In this CD, you will see many remote sensing images around Asia acquired by earth observation satellites. These remote sensing satellites are equipped with sensors looking down to the earth. They are the "eyes in the sky" constantly observing the earth as they go round in predictable orbits.

Effects of Atmosphere

In satellite remote sensing of the earth, the sensors are looking through a layer of atmosphere separating the sensors from the Earth's surface being observed. Hence, it is essential to understand the effects of atmosphere on the electromagnetic radiation travelling from the Earth to the sensor through the atmosphere. The atmospheric constituents cause wavelength dependent absorption and scattering of radiation. These effects degrade the quality of images. Some of the atmospheric effects can be corrected before the images are subjected to further analysis and interpretation.

A consequence of atmospheric absorption is that certain wavelength bands in the electromagnetic spectrum are strongly absorbed and effectively blocked by the atmosphere. The wavelength regions in the electromagnetic spectrum usable for remote sensing are determined by their ability to penetrate atmosphere. These regions are known as the atmospheric transmission windows. Remote sensing systems are often designed to operate within one or more of the atmospheric windows. These windows exist in the microwave region, some wavelength bands in the infrared, the entire visible region and part of the near ultraviolet regions. Although the atmosphere is practically transparent to x-rays and gamma rays, these radiations are not normally used in remote sensing of the earth.

Optical and Infrared Remote Sensing

In Optical Remote Sensing, optical sensors detect solar radiation reflected or scattered from the earth, forming images resembling photographs taken by a camera high up in space. The wavelength region usually extends from the visible and near infrared (commonly abbreviated as VNIR) to the short-wave infrared (SWIR).

Different materials such as water, soil, vegetation, buildings and roads reflect visible and infrared light in different ways. They have different colours and brightness when seen under the sun. The interpretation of optical images require the knowledge of the spectral reflectance signatures of the various materials (natural or man-made) covering the surface of the earth.

There are also infrared sensors measuring the thermal infrared radiation emitted from the earth, from which the land or sea surface temperature can be derived.

Microwave Remote Sensing

There are some remote sensing satellites which carry passive or active microwave sensors. The active sensors emit pulses of microwave radiation to illuminate the areas to be imaged. Images of the earth surface are formed by measuring the microwave energy scattered by the ground or sea back to the sensors. These satellites carry their own "flashlight" emitting microwaves to illuminate their targets. The images can thus be acquired day and night. Microwaves have an additional advantage as they can penetrate clouds. Images can be acquired even when there are clouds covering the earth surface.

A microwave imaging system which can produce high resolution image of the Earth is the synthetic aperture radar (SAR). The intensity in a SAR image depends on the amount of microwave backscattered by the target and received by the SAR antenna. Since the physical mechanisms responsible for this backscatter is different for microwave, compared to visible/infrared radiation, the interpretation of SAR images requires the knowledge of how microwaves interact with the targets.

REMOTE SENSING IMAGES

Remote sensing images are normally in the form of digital images. In order to extract useful information from the images, image processing techniques may be employed to enhance the image to help visual interpretation, and to correct or restore the image if the image has been subjected to geometric distortion, blurring or degradation by other factors. There are many image analysis techniques available and the methods used depend on the requirements of the specific problem concerned. In many cases, image segmentation and classification algorithms are used to delineate different areas in an image into thematic classes. The resulting product is a thematic map of the study area. This thematic map can be combined with other databases of the test area for further analysis and utilization.

Visual System

Passive Remote Sensing: The eyes passively senses the radiation reflected or emitted from the object. The sensing system depends on an external source of illumination.

The human visual system is an example of a remote sensing system in the general sense. The sensors in this example are the two types of photosensitive cells, known as the cones and the rods, at the retina of the eyes. The cones are responsible for colour vision. There are three types of cones, each being sensitive to one of the red, green, and blue regions of the visible spectrum. Thus, it is not coincidental that the modern computer display monitors make use of the same three primary colours to generate a multitude of colours for displaying colour images. The cones are insensitive under low light illumination condition, when their jobs are taken over by the rods. The rods are sensitive only to the total light intensity. Hence, everything appears in shades of grey when there is insufficient light.

As the objects/events being observed are located far away from the eyes, the information needs a carrier to travel from the object to the eyes. In this case, the information carrier is the visible light, a part of the electromagnetic spectrum.

The objects reflect/scatter the ambient light falling onto them. Part of the scattered light is intercepted by the eyes, forming an image on the retina after passing through the optical system of the eyes. The signals generated at the retina are carried via the nerve fibres to the brain, the central processing unit (CPU) of the visual system.

These signals are processed and interpreted at the brain, with the aid of previous experiences. When operating in this mode, the visual system is an example of a "Passive Remote Sensing" system which depends on an external source of energy to operate.

We all know that this system won't work in darkness. However, we can still see at night if we provide our own source of illumination by carrying a flashlight and shining the beam towards the object we want to observe. In this case, we are performing "Active Remote Sensing", by supplying our own source of energy for illuminating the objects.

The Planet Earth

The planet Earth is the third planet in the solar system located at a mean distance of about 1.50 x 10^8 km from the sun, with a mass of 5.97 x 10^{24} kg. Descriptions of the shape of the earth have evolved from the flat-earth model, spherical model to the currently accepted ellipsoidal model derived from accurate ground surveying and satellite measurements.

A number of reference ellipsoids have been defined for use in identifying the three dimensional coordinates (*i.e.* position in space) of a point on or above the earth surface for the purpose of surveying, mapping and navigation. The reference ellipsoid in the World Geodetic System 1984 (WGS-84) commonly used in satellite Global Positioning System (GPS) has the following parameters:

- Equatorial Radius = 6378.1370 km
- Polar Radius = 6356.7523 km

The earth's crust is the outermost layer of the earth's land surface. About 29.1 per cent of the earth's crust area is above sea level. The rest is covered by water. A layer of gaseous atmosphere envelopes the earth's surface.

The Earth's Atmosphere

The earth's surface is covered by a layer of atmosphere consisting of a mixture of gases and other solid and liquid particles. The gaseous materials extend to several hundred kilometers in altitude, though there is no well defined boundary for the upper limit of the atmosphere. The first 80 km of the atmosphere contains more than 99 per cent of the total mass of the earth's atmosphere.

Vertical Structure of the Atmosphere

The vertical profile of the atmosphere is divided into four layers: troposphere, stratosphere, mesosphere and thermosphere. The tops of these layers are known as the tropopause, stratopause, mesopause and thermopause, respectively.

- Troposphere: This layer is characterized by a decrease in temperature with respect to height, at a rate of about 6.5°C per kilometer, up to a height of about 10 km. All the weather activities (water vapour, clouds, precipitation) are confined to this layer. A layer of aerosol particles normally exists near to the earth surface. The aerosol concentration decreases nearly exponentially with height, with a characteristic height of about 2 km.
- Stratosphere: The temperature at the lower 20 km of the stratosphere is approximately constant, after which the temperature increases with height, up to an altitude of about 50 km. Ozone exists mainly at the stratopause. The troposphere and the stratosphere together account for more than 99 per cent of the total mass of the atmosphere.
- Mesosphere: The temperature decreases in this layer from an altitude of about 50 km to 85 km.
- Thermosphere: This layer extends from about 85 km upward to several hundred kilometers. The temperature may range from 500 K to 2000 K. The gases exist mainly in the form of thin plasma, *i.e.* they are ionized due to bombardment by solar ultraviolet radiation and energetic cosmic rays.

The term upper atmosphere usually refers to the region of the atmosphere above the troposphere.

Many remote sensing satellites follow the near polar sun-synchronous orbits at a height around 800 km, which is well above the thermopause.

Atmospheric Constituents

The atmosphere consists of the following components:

- Permanent Gases: They are gases present in nearly constant concentration, with little spatial variation. About 78 per cent by volume of the atmosphere is nitrogen while the life-sustaining oxygen occupies 21 per cent. The remaining one per cent consists of the inert gases, carbon dioxide and other gases.
- Gases with Variable Concentration: The concentration of these gases may vary greatly over space and time. They consist of water vapour, ozone, nitrogeneous and sulphurous compounds.
- Solid and liquid particulates: Other than the gases, the atmosphere also contains solid and liquid particles such as aerosols, water droplets and ice crystals. These particles may congregate to form clouds and haze.

Electromagnetic Waves

Electromagnetic waves are energy transported through space in the form of periodic disturbances of electric and magnetic fields. All electromagnetic waves travel through space at the same speed, c = 2.99792458 x 10^8 m/s, commonly known as the speed of light. An electromagnetic wave is characterized by a frequency and a wavelength. These two quantities are related to the speed of light by the equation:

Speed of light = frequency x wavelength

The frequency (and hence, the wavelength) of an electromagnetic wave depends on its source.

There is a wide range of frequency encountered in our physical world, ranging from the low frequency of the electric waves generated by the power transmission lines to the very high frequency of the gamma rays originating from the atomic nuclei. This wide frequency range of electromagnetic waves constitute the Electromagnetic Spectrum.

The Electromagnetic Spectrum

The electromagnetic spectrum can be divided into several wavelength (frequency) regions, among which only a narrow band from about 400 to 700 nm is visible to the human eyes. Note that there is no sharp boundary between these regions. The boundaries shown in the above figures are approximate and there are overlaps between two adjacent regions.

Wavelength units: 1 mm = 1000 μm; 1 μm = 1000 nm.

- Radio Waves: 10 cm to 10 km wavelength.
- Microwaves: 1 mm to 1 m wavelength. The microwaves are further divided into different frequency (wavelength) bands: (1 GHz = 10^9 Hz)
  - P band: 0.3-1 GHz (30-100 cm)
  - L band: 1-2 GHz (15-30 cm)
  - S band: 2-4 GHz (7.5-15 cm)
  - C band: 4-8 GHz (3.8-7.5 cm)
  - X band: 8-12.5 GHz (2.4-3.8 cm)
  - Ku band: 12.5-18 GHz (1.7-2.4 cm)
  - K band: 18-26.5 GHz (1.1-1.7 cm)
  - Ka band: 26.5-40 GHz (0.75-1.1 cm)
- Infrared: 0.7 to 300 μm wavelength. This region is further divided into the following bands:
  - Near Infrared (NIR): 0.7 to 1.5 μm.
  - Short Wavelength Infrared (SWIR): 1.5 to 3 μm.
  - Mid Wavelength Infrared (MWIR): 3 to 8 μm.
  - Long Wanelength Infrared (LWIR): 8 to 15 μm.
  - Far Infrared (FIR): longer than 15 μm.

The NIR and SWIR are also known as the Reflected Infrared, referring to the main infrared component of the solar radiation reflected from the earth's surface. The MWIR and LWIR are the Thermal Infrared.

- Visible Light: This narrow band of electromagnetic radiation extends from about 400 nm (violet) to about 700 nm (red). The various colour components of the visible spectrum fall roughly within the following wavelength regions:
  - Red: 610-700 nm
  - Orange: 590-610 nm
  - Yellow: 570-590 nm
  - Green: 500-570 nm
  - Blue: 450-500 nm
  - Indigo: 430-450 nm
  - Violet: 400-430 nm
- Ultraviolet: 3 to 400 nm
- X-Rays and Gamma Rays

SIGNATURE IN REMOTE SENSING

The knowledge of spectral signatures is essential for exploiting the potential of remote sensing techniques. This knowledge enables one to identify and classify the objects of agricultural resources. It is also required for interpretation of all remotely sensed data, especially in agricultural resource data whether the interpretation is carried out visually or using digital techniques.

It also helps us in specifying requirements for any remote sensing mission *e.g.* which optimal wave length bands to be used or which type of sensor will be best suited for a particular task (agricultural survey).

All objects of agricultural resource on the surface of the earth have characteristic spectral signatures. The average spectral reflectance curves (or) spectral signatures for three typical earth's features; vegetation, soil and water. The spectral reflectance curves for vigorous vegetation manifests the "Peak-and valley" configuration.

The valleys in the visible portion of the spectrum are indicative of pigments in plant leaves. Dips in reflectance that can be sent at wavelengths of 0.65 mm, 1.4 mm and 1.9 mm are attributable to absorption of water by leaves. The soil curves show a more regular variation of reflectance.

Factors that evidently affect soil reflectance are moisture content, soil texture, surface roughness and presence of organic matter. The water curves shows that from about 0.5 mm, reduction in reflectance with increasing wavelength, so that in the near infrared range, the reflectance of deep clear water is virtually zero (Mather, 1987) However, the spectral reflectance of water is significantly affected by the presence of dissolved and suspended organic and inorganic material and by the depth of the water body.

Determinations of spectral signatures implies basic under standing of interaction of electromagnetic radiation with agricultural resources objects. This is also necessary for analysing and designing sensor systems for agricultural survey.

SENSOR SYSTEMS IN REMOTE SENSING

In remote sensing the acquisition of data is depending upon the sensor system used. Various remote sensing platforms (Aircraft, Satellite) are equipped with different sensor systems. Sensor is a device that receives electromagnetic radiation, converts it into a signal and presents it in a form suitable of obtaining information about the land or earth resource as used by an information gathering system. Sensor can be grouped, either on the basic of energy source. They are as classified.

Active sensor: An active sensor operates by emitting its own energy, which is needed to detect the various phenomena (*e.g.* RADAR, camera with a flash gun).

Passive sensor: The operation of passive sensor is dependent of the existing sources of energy, like sun (*e.g.* photographic systems, multispectral scanners).

The given sensor system of camera are in agricultural survey.

Photographic Cameras

The photographic system, having conventional camera with black and white photography, is the oldest and probably, so far, the most widely used sensor for recording information about ground object. Photographic cameras have been successfully used in aircraft platform remote sensing.

In this system, the information is limited to size and shape, as the films used are sensitive only to visible region of spectrum. The response of black and white films is about 0.4-0.7 mm for infrared imagery, films with response extending up to 0.9 mm are available.

Return Beam Vidicon (RBM)

This is very similar to a television camera. In such a system, a fixed camera lens on a photosensitive semi-transparent sheet forms the ground image. This image is created on the surface as electrical change or potential. The TV cameras are the best example of high resolution, operated in space for resource survey was the RBV used in LAND SAT series.

On LAND SAT I, II and III RBV cameras were used, each corresponding to a different wavelength band 0.475-0.585 mm (green), 0.580-0.690 mm (red) and 0.690-0.830 mm (near infrared).

The Indian experimental remote sensing satellite, Bhaskara-I and II carried a two-band TV camera system, Multispectral imagery was produced in LAND SAT and Bhaskara by using separate camera tubes of each band and selecting the spectral band with appropriate filters.

Optical-Mechanical Scanners

This imaging system has the advantage that any set of desired spectral bands can be selected with appropriate filter and detector combinations. The mostly widely used sensor in this category is the MSS on LAND SAT series. MSS has four spectral bands, covering form 0.5-to 1.1 mm region. MSS operates on the principle of scanning successive lines at right angles to the flight path by means of a rotation or oscillating optical system.

The radiation levels along the lines are recorded by appropriate sensor elements. When used in the visible band, the collected light can be split by the optics and separately filtered and recorded, giving simultaneous multispectral recording from the one instrument. MSS can record in any part of ultraviolet to near IR window. They are use also in the thermal IR windows.

Radar and Microwave Sensors

The acquisition of data in microwave region has been possible since 1950s but its application to natural resources is considerably less developed, as compared to the visible and IR image interpretations.

Microwave sensors have distinct advantages because they are unaffected by atmospheric conditions and are thus able to penetrated smoke, clouds, haze and snow.

Under this system, Plan Position Indicator (PPI), Side Looking Air borne Radar (SLAR) and Synthetic Aperture Radar (SAR) can be grouped. These systems offer day and night as well as all weather capability and ability to penetrate a cover of vegetation.

Advance Remote Sensors

Linear Imaging and Self Scanning Sensors (LISS) are the advanced imaging systems. This type of scanning sensor are used an array of solid-state devices. The array may be made of photo-diodes, phototransistors or Charge-Coupled Devices (CCDs). In the LISS, the optics focuses a strip of terrain in the cross-track into the sensor array.

The image from each detector is stored and shifted out sequentially to receive a video signal. The SPOT (Satellite Probatorie d' Observation de la Terra) and IRS (Indian Remote Sensing Satellite) series carry such solid-state sensor systems, which are also known as push-broom scanners.

The IRS IC most advanced satellite, carries an improved sensor system. Besides carrying a sophisticated LISS-III camera, it has a Panchromatic camera (PAN) and a Wide Field Sensor (WiFS).

The PAN has been designed to provide data with a spatial resolution of 5.8m in stereo mode, with a ground swath of 70km, whereas WiFS provides data in two spectral bands, with a spectral resolution of 188m and a ground swath of 180km.

DATA ACQUISITION TECHNIQUES IN REMOTE SENSING

Remote sensing is the acquisition of information about an object or phenomenon, without making physical contact with the object. In modern usage, the term generally refers to the use of aerial sensor technologies to detect and classify objects on Earth (both on the surface, and in the atmosphere and oceans) by means of propagated signals (*e.g.* electromagnetic radiation emitted from aircraft or satellites).

There are two main types of remote sensing: passive remote sensing and active remote sensing. Passive sensors detect natural radiation that is emitted or reflected by the object or surrounding area being observed. Reflected sunlight is the most common source of radiation measured by passive sensors. Examples of passive remote sensors include film photography, infrared, charge-coupled devices, and radiometers.

Active collection, on the other hand, emits energy in order to scan objects and areas whereupon a sensor then detects and measures the radiation that is reflected or backscattered from the target.

RADAR is an example of active remote sensing where the time delay between emission and return is measured, establishing the location, height, speed and direction of an object.

Remote sensing makes it possible to collect data on dangerous or inaccessible areas.

Remote sensing applications include monitoring deforestation in areas such as the Amazon Basin, glacial features in Arctic and Antarctic regions, and depth sounding of coastal and ocean depths.

Military collection during the cold war made use of stand-off collection of data about dangerous border areas. Remote sensing also replaces costly and slow data collection on the ground, ensuring in the process that areas or objects are not disturbed.

Orbital platforms collect and transmit data from different parts of the electromagnetic spectrum, which in conjunction with larger scale aerial or ground-based sensing and analysis, provides researchers with enough information to monitor trends such as El Niño and other natural long and short term phenomena. Other uses include different areas of the earth sciences such as natural resource management, agricultural fields such as land usage and conservation, and national security and overhead, ground-based and stand-off collection on border areas.

By satellite, aircraft, spacecraft, buoy, ship, and helicopter images, data is created to analyze and compare things like vegetation rates, erosion, pollution, forestry, weather, and land use. These things can be mapped, imaged, tracked and observed. The process of remote sensing is also helpful for city planning, archaeological investigations, military observation and geomorphological surveying.

Data Acquisition Techniques

The basis for multispectral collection and analysis is that of examined areas or objects that reflect or emit radiation that stand out from surrounding areas.

APPLICATIONS OF REMOTE SENSING DATA

- Conventional radar is mostly associated with aerial traffic control, early warning, and certain large scale meteorological data. Doppler radar is used by local law enforcements' monitoring of speed limits and in enhanced meteorological collection such as wind speed and direction within weather systems. Other types of active collection includes plasmas in the ionosphere. Interferometric synthetic aperture radar is used to produce precise digital elevation models of large scale terrain.
- Laser and radar altimeters on satellites have provided a wide range of data. By measuring the bulges of water caused by gravity, they map features on the seafloor to a resolution of a mile or so. By measuring the height and wave-length of ocean waves, the altimeters measure wind speeds and direction, and surface ocean currents and directions.
- Light detection and ranging (LIDAR) is well known in examples of weapon ranging, laser illuminated homing of projectiles. LIDAR is used to detect and measure the concentration of various chemicals in the atmosphere, while airborne LIDAR can be used to measure heights of objects and features on the ground more accurately than with radar technology. Vegetation remote sensing is a principal application of LIDAR.
- Radiometers and photometers are the most common instrument in use, collecting reflected and emitted radiation in a wide range of frequencies. The most common are visible and infrared sensors, followed by microwave, gamma ray and rarely, ultraviolet. They may also be used to detect the emission spectra of various chemicals, providing data on chemical concentrations in the atmosphere.
- Stereographic pairs of aerial photographs have often been used to make topographic maps by imagery and terrain analysts in trafficability and highway departments for potential routes.
- Simultaneous multi-spectral platforms such as Landsat have been in use since the 70's. These thematic mappers take images in multiple wavelengths of electro-magnetic radiation (multi-spectral) and are usually found on Earth observation satellites, including (for example) the Landsat programme or the IKONOS satellite. Maps of land cover and land use from thematic mapping can be used to prospect for

minerals, detect or monitor land usage, deforestation, and examine the health of indigenous plants and crops, including entire farming regions or forests.

- Hyperspectral imaging produces an image where each pixel has full spectral information with imaging narrow spectral bands over a contiguous spectral range. Hyperspectral imagers are used in various applications including mineralogy, biology, defence, and environmental measurements.
- Within the scope of the combat against desertification, remote sensing allows to follow-up and monitor risk areas in the long term, to determine desertification factors, to support decision-makers in defining relevant measures of environmental management, and to assess their impacts.

Geodetic

- Overhead geodetic collection was first used in aerial submarine detection and gravitational data used in military maps. This data revealed minute perturbations in the Earth's gravitational field (geodesy) that may be used to determine changes in the mass distribution of the Earth, which in turn may be used for geological studies.

Acoustic and Near-Acoustic

- Sonar: *passive sonar*, listening for the sound made by another object (a vessel, a whale etc); *active sonar*, emitting pulses of sounds and listening for echoes, used for detecting, ranging and measurements of underwater objects and terrain.
- Seismograms taken at different locations can locate and measure earthquakes (after they occur) by comparing the relative intensity and precise timing.

To coordinate a series of large-scale observations, most sensing systems depend on the following: platform location, what time it is, and the rotation and orientation of the sensor. High-end instruments now often use positional information from satellite navigation systems.

The rotation and orientation is often provided within a degree or two with electronic compasses. Compasses can measure not just azimuth (i. e. degrees to magnetic north), but also altitude (degrees above the horizon), since the magnetic field curves into the Earth at different angles at different latitudes. More exact orientations require gyroscopic-aided orientation, periodically realigned by different methods including navigation from stars or known benchmarks. Resolution impacts collection and is best explained with the following relationship: less resolution=less detail and larger coverage, More resolution=more detail, less coverage.

The skilled management of collection results in cost-effective collection and avoid situations such as the use of multiple high resolution data which tends to clog transmission and storage infrastructure.

Data Processing

Generally speaking, remote sensing works on the principle of the *inverse problem*. While the object or phenomenon of interest (the state) may not be directly measured, there exists some other variable that can be detected and measured (the observation), which may be related to the object of interest through the use of a data-derived computer model.

The common analogy given to describe this is trying to determine the type of animal from its footprints. For example, while it is impossible to directly measure temperatures in the upper atmosphere, it is possible to measure the spectral emissions from a known chemical species (such as carbon dioxide) in that region.

The frequency of the emission may then be related to the temperature in that region via various thermodynamic relations.

The quality of remote sensing data consists of its spatial, spectral, radiometric and temporal resolutions.

Spatial Resolution

The size of a pixel that is recorded in a raster image – typically pixels may correspond to square areas ranging in side length from 1 to 1,000 metres (3.3 to 3,300 ft).

Spectral Resolution

The wavelength width of the different frequency bands recorded – usually, this is related to the number of frequency bands recorded by the platform. Current Landsat collection is that of seven bands, including several in the infra-red spectrum, ranging from a spectral resolution of 0.07 to 2.1 ìm. The Hyperion sensor on Earth Observing-1 resolves 220 bands from 0.4 to 2.5 ìm, with a spectral resolution of 0.10 to 0.11 ìm per band.

Radiometric Resolution

The number of different intensities of radiation the sensor is able to distinguish. Typically, this ranges from 8 to 14 bits, corresponding to 256 levels of the gray scale and up to 16,384 intensities or "shades" of colour, in each band. It also depends on the instrument noise.

Temporal Resolution

The frequency of flyovers by the satellite or plane, and is only relevant in time-series studies or those requiring an averaged or mosaic image as in deforesting monitoring.

This was first used by the intelligence community where repeated coverage revealed changes in infrastructure, the deployment of units or the modification/ introduction of equipment. Cloud cover over a given area or object makes it necessary to repeat the collection of said location.

In order to create sensor-based maps, most remote sensing systems expect to extrapolate sensor data in relation to a reference point including distances between known points on the ground.

This depends on the type of sensor used. For example, in conventional photographs, distances are accurate in the centre of the image, with the distortion of measurements increasing the farther you get from the centre. Another factor is that of the platen against which the film is pressed can cause severe errors when photographs are used to measure ground distances. The step in which this problem is resolved is called georeferencing, and involves computer-aided matching up of points in the image (typically 30 or more points per image) which is extrapolated with the use of an established benchmark, "warping" the image to produce accurate spatial data. As of the early 1990s, most satellite images are sold fully georeferenced. In addition, images may need to be radiometrically and atmospherically corrected.

Radiometric Correction

Gives a scale to the pixel values, *e.g.* the monochromatic scale of 0 to 255 will be converted to actual radiance values.

Atmospheric correction eliminates atmospheric haze by rescaling each frequency band so that its minimum value (usually realised in water bodies) corresponds to a pixel value of 0.

The digitizing of data also make possible to manipulate the data by changing gray-scale values.

Interpretation is the critical process of making sense of the data. The first application was that of aerial photographic collection which used the following process; spatial measurement through the use of a light table in both conventional single or stereographic coverage, added skills such as the use of photogrammetry, the use of photomosaics, repeat coverage, Making use of objects' known dimensions in order to detect modifications. Image Analysis is the recently developed automated computer-aided application which is in increasing use.

Object-Based Image Analysis (OBIA) is a sub-discipline of GIScience devoted to partitioning remote sensing (RS) imagery into meaningful image-objects, and assessing their characteristics through spatial, spectral and temporal scale.

Old data from remote sensing is often valuable because it may provide the only long-term data for a large extent of geography. At the same time, the data is often complex to interpret, and bulky to store. Modern systems tend to store the data digitally, often with lossless compression. The difficulty with this approach is that the data is fragile, the format may be archaic, and the data may be easy to falsify.

One of the best systems for archiving data series is as computer-generated machine-readable ultrafiche, usually in typefonts such as OCR-B, or as digitized half-tone images. Ultrafiches survive well in standard libraries, with lifetimes of several centuries.

They can be created, copied, filed and retrieved by automated systems. They are about as compact as archival magnetic media, and yet can be read by human beings with minimal, standardized equipment.

Data Processing Levels

To facilitate the discussion of data processing in practice, several processing "levels" were first defined in 1986 by NASA as part of its Earth Observing System and steadily adopted since then, both internally at NASA and elsewhere; these definitions are:

Level	Description
0	Reconstructed, unprocessed instrument and payload data at full resolution, with any and all communications artifacts (*e.g.*, synchronization frames, communications headers, duplicate data) removed.
1a	Reconstructed, unprocessed instrument data at full resolution, time-referenced, and annotated with ancillary information, including radiometric and geometric calibration coefficients and georeferencing parameters (*e.g.*, platform ephemeris) computed and appended but not applied to the Level 0 data (or if applied, in a manner that level 0 is fully recoverable from level 1a data).
1b	Level 1a data that have been processed to sensor units (*e.g.*, radar backscatter cross section, brightness temperature, etc.); not all instruments have Level 1b data; level 0 data is not recoverable from level 1b data.
2	Derived geophysical variables (*e.g.*, ocean wave height, soil moisture, ice concentration) at the same resolution and location as Level 1 source data.
3	Variables mapped on uniform spacetime grid scales, usually with some completeness and consistency (*e.g.*, missing points interpolated, complete regions mosaicked together from multiple orbits, etc).
4	Model output or results from analyses of lower level data (*i.e.*, variables that were not measured by the instruments but instead are derived from these measurements).

A Level 1 data record is the most fundamental (i. e., highest reversible level) data record that has significant scientific utility, and is the foundation upon which all subsequent data sets are produced. Level 2 is the first level that is directly usable for most scientific applications; its value is much greater than the lower levels. Level 2 data sets tend to be less voluminous than Level 1 data because they have been reduced temporally, spatially, or spectrally. Level 3 data sets are generally smaller than lower level data sets and thus can be dealt with without incurring a great deal of data handling overhead. These data tend to be generally more useful for many applications. The regular spatial and temporal organization of Level 3 datasets makes it feasible to readily combine data from different sources.

History

The modern discipline of remote sensing arose with the development of flight. The balloonist G. Tournachon (alias Nadar) made photographs of Paris from his balloon in 1858. Messenger pigeons, kites, rockets and unmanned balloons were also used for early images. With the exception of balloons, these first, individual images were not particularly useful for map making or for scientific purposes. Systematic aerial photography was developed for military surveillance and reconnaissance purposes beginning in World War I and reaching a climax during the Cold War with the use of modified combat aircraft such as the P-51, P-38, RB-66 and the F-4C, or specifically designed collection platforms such as the U2/TR-1, SR-71, A-5 and the OV-1 series both in overhead and stand-off collection. A more recent development is that of increasingly smaller sensor pods such as those used by law enforcement and the military, in both manned and unmanned platforms.

The advantage of this approach is that this requires minimal modification to a given airframe. Later imaging technologies would include Infra-red, conventional, doppler and synthetic aperture radar.

The development of artificial satellites in the latter half of the 20th century allowed remote sensing to progress to a global scale as of the end of the Cold War. Instrumentation aboard various Earth observing and weather satellites such as Landsat, the Nimbus and more recent missions such as RADARSAT and UARS provided global measurements of various data for civil, research, and military purposes. Space probes to other planets have also provided the opportunity to conduct remote sensing studies in extraterrestrial environments, synthetic aperture radar aboard the Magellan spacecraft provided detailed topographic maps of Venus, while instruments aboard SOHO allowed studies to be performed on the Sun and the solar wind, just to name a few examples.

Recent developments include, beginning in the 1960s and 1970s with the development of image processing of satellite imagery. Several research groups in Silicon Valley including NASA Ames Research Centre, GTE and ESL Inc.

developed Fourier transform techniques leading to the first notable enhancement of imagery data.

ACQUIRING REMOTE SENSING DATA

Remote sensors can be grouped according to the number of bands and the frequency range of those bands that the sensor can detect. Common categories of remote sensors include *panchromatic*, *multispectral*, *hyperspectral*, and *ultraspectral* sensors.

Panchromatic sensors cover a wide band of wavelengths in the visible light or near infrared light spectrum. An example of a single band sensor of this type would be a black and white photographic film camera.

Multispectral sensors cover two or more spectral bands simultaneously typically from 0.3 m to 14 m wide.

Hyperspectral sensors cover spectral bands narrower than multispectral sensors. Image data from several hundred bands are recorded at the same time offering much greater spectral resolution than a sensor covering wider bands.

Ultrasprectral sensors are still under development and not yet in use. These sensors will cover thousands of bands with an even narrower bandwidth than hyperspectral sensors.

DIGITAL IMAGE DATA DELIVERY SYSTEMS

Electro-optical and spectral imaging scanners produce digital images with the use of detectors that measure the brightness of reflected electromagnetic energy. Scanners consist of one or more sensor detectors depending on type of sensor system used. One type of scanner is called a whiskbroom scanner also referred to as across-track scanners. It uses rotating mirrors to scan the landscape below from side to side perpendicular to the direction of the sensor platform, like a whiskbroom. The width of the sweep is referred to as the sensor swath.

The rotating mirrors redirect the reflected light to a point where a single or just a few sensor detectors are grouped together. Whiskbroom scanners with their moving mirrors tend to be large and complex to build. The moving mirrors create spatial distortions that must be corrected with preprocessing by the data provider before image data is delivered to the user. An advantage of whiskbroom scanners is that they have fewer sensor detectors to keep calibrated as compared to other types of sensors.

Another type of scanner, which does not use rotating mirrors, is the pushbroom scanner also referred to as an along-track scanner. The sensor detectors in a pushbroom scanner are lined up in a row called a linear array. Instead of sweeping from side to side as the sensor system moves forward, the one dimensional sensor array captures the entire scan line at once like a pushbroom would. Some recent scanners referred to as step stare scanners contain two-dimensional arrays in rows and columns for each band. Pushbroom scanners are lighter, smaller and less complex because of fewer moving parts than whiskbroom scanners. Also they have better radiometric and spatial resolution. A major disadvantage of pushbroom scanners is the calibration required for a large number of detectors that make up the sensor system.

DATA PROCESSING, INTERPRETATION AND ANALYSIS

Much of the technical work of remote sensing involves pre-processing and applying radiometric and geometric corrections to imagery to compensate for errors due to factors such as atmospheric interference of incoming radiation and sensor and data stream irregularities. Once such corrections are applied, imagery must be *georeferenced*. Georeferencing is the process of taking the image in its raw format (rows and columns of data) and linking it to the land that it covers. Images are georeferenced by linking spatially distributed control points in the satellite image to points on base maps or points referenced in the field through global positioning systems. The raster data in the image is thereby registered to a Cartesian coordinate system, and can be combined with other georeferenced data sets in a geographic information system.

The processed data can now either be visually interpreted or classified using manual or automated processes.

The main elements of visual image interpretation involve gradients of tone or colour, resolution, size and shape, texture and pattern, site and association, and height and shadows. Given their knowledge of the characteristic spectral signatures of different land cover types, scientists my inspect black and white images of each band separately in order to identify features and patterns.

For many purposes, data that is collected from the earth's surface, which represents a continuous variation, needs to be categorized (de Sherbinin *et al.* 2002). Image classification is the process of creating discrete classes or categories of land cover, utilizing information from some or all of the bands to group together pixels with similar spectral signatures.

Supervised classification entails providing the software with sample pixels that represent specific features, such as boreal forest, and then having the computer classify every pixel with a similar *spectral signature* as boreal forest.

Analysts may also use images from different seasons in order to discriminate vegetation cover types that have different phenologies, such as deciduous and evergreen forests. In unsupervised classification, the analyst specifies the desired number of classes, and the computer automatically sorts the pixels according to their spectral signatures. The analyst then labels the resulting groups based on some local knowledge of the land cover patterns.

Once classified, it is necessary to verify or validate that the output product accurately represents the actual composition, content, structure or land surface characteristics being mapped. Validation requires either field visits, ground-truthing or comparing the classified image with existing maps or images of sufficient detail. Statistics can be derived for the classified imagery indicating the general and specific (class-wise) agreement between the pixels or classes used, letting the user know which were classified correctly and which ones were not.

Validation results are also sometimes presented as a percentage value associated with the map that communicates how accurate the map is on a per pixel basis. Since the highest confidence rankings reported by satellite land cover data sets are between 85 per cent and 90 per cent (for the easiest types of land cover to classify), for an image with a per pixel accuracy of 85 per cent the probability that one pixel out of four is incorrectly classified is close to 0.50. The output of remote sensing data analysis can be presented in a variety of ways including a printout of the enhanced image itself a spatial database, summary statistics and/or graphs. The output data can be integrated with a geographic information system (GIS) database for further analysis.

SENSORS CHARACTERISTICS

A sensor is characterized by its spectral properties (number and placement of bands), its orbital altitude and path, its swath width, and its *spatial resolution*. Spatial resolution is measured in terms of the size of one pixel projected on the ground. Spatial resolution is directly tied to the size of the features that can be resolved (or "seen") on the ground. The higher the resolution, the less likely that there will be "mixed pixels" in which radiances effectively represent an average of land cover types in the ground area represented by that pixel (*e.g.*, half lake and half forest). Commercial high resolution sensors have a spatial resolution in the 0.6-10 meter range, medium resolution sensors fall in the 10-50 meter range, and low resolution sensors have greater than 50 meter resolution.

Until the advent of the commercial satellites IKONOS and QuickBird, with resolutions of one square meter or finer, high resolution imagery was the exclusive province of intelligence-gathering agencies. Most social science applications do not command the financial resources required to obtain such high resolution data, nor are images of this resolution generally required, except perhaps in the area of international relations, law and policy. Most social science research tends to utilize polar-orbiting satellites with medium spatial resolution, such as the Landsat, SPOT, TERRA and, more recently, AQUA satellites. They provide good spectral and ground resolution, with multiple visible, infrared, and panchromatic bands and pixel width ranging from 5 to 30 meters.

PLATFORMS AND ORBITS

High-altitude remote sensing originated in the mid-1800s with aerial photography by balloon and, in at least one instance, the use of cameras attached to the underside of birds. Airplanes became the dominant remote sensing "platform" by the early 20th century. This practice continues to evolve for wartime, intelligence, commercial and government applications. An advantage of airborne remote sensing, is the capability of offering very high spatial resolution images (20 cm or less). The disadvantages are low coverage area and high cost per unit area of ground coverage. It is not cost-effective to map a large area using an airborne remote sensing system. Airborne remote sensing missions are often carried out as one-time operations, whereas earth observation satellites offer the possibility of continuous monitoring of the earth. The development of satellite remote sensing had greatly improved the ability to cover large areas.

There are two groups of satellites depending on the orbit in which they are placed. A *geostationary* orbit is established when a satellite is placed at a very high altitude, roughly 36,000 km above the earth's equator, and caused to orbit with the earth's rotation (called a *prograde* orbit). The altitude may vary slightly from one geostationary satellite to the next, depending upon the mass of the satellite, but, for the most part, this is a fundamental physical constraint. The rules of geometry-that is, the sight line from the satellite's position above the equator to the farthest edge of the earth's sphere-dictate that geostationary satellites can only "see" limited area at any one time. The laws of physics and capabilities of engineering limit their spatial resolution to a range of about 1 to 10 square kilometres. The total image or scene size, known as the *field of view*, is often thousands of kilometres. Thus, unless a geostationary satellite spins or turns its optics, its view is necessarily fixed. This allows for continuous monitoring, and often a very large, synoptic view of much of one entire hemisphere. The coarse (km range) resolution versus the wide, continuous field of view constitute the main tradeoffs to consider for this orbit type. These characteristics make geostationary satellites best for collecting weather and climate data (such as cloud cover and surface temperature) and relaying communications data, although AVHRR data are used for global-and regional-scale land cover analyses.

The other group of satellites, by far the largest group of earth-orbiting satellites is with the *sun-synchronous* or *polar-orbiting*. These are lunched below the altitude of geostationary satellites closer to the earth's surface, at orbits ranging from 700km to 1000km. These satellites usually orbit at a steep inclination relative to the equator, in the direction opposite the earth's rotation, known as a *retrograde* orbit. When the satellite's orbit and earth's rotation are combined, they result in an s-shaped path relative to a map of the earth's surface. Given enough time, the orbits and rotations of the earth bring the satellite over the same location, leading to the term *exact repeat* satellites. The number of orbits between each return to the same longitude and latitude is called the *repeat cycle*. These satellites usually orbit the earth in roughly 100 to 120 minutes, circling several times per day, returning a satellite to the same position over the earth's surface only after 2 weeks or more. The speed of motion limits the time that a satellite spends over a location, and the amount of time a scanner can "look" at any single ground cell (called the *dwell time*). Most exact repeat satellites that use passive sensors are also in sun synchronous orbits, meaning that they cross the same latitude at the same daylight time with each orbit, but with their location shifted to a different longitude. Their lower altitude allows these satellites to obtain images with spatial resolution ranging from 1-200 meters per side of a pixel, and an image width ranging from tens to thousands of kilometres per scene.

As the satellite passes over the earth's surface, its motion can be described in terms of a ground track that it follows at a certain altitude. Most satellites are *nadir looking*, meaning that their sensing equipment is aimed straight down towards the centre of the earth. However, normal measurements generally include areas substantially on either side of this ground track, and that total area is called the *swath width*. Because most low-orbiting satellites follow a polar orbit, their ground tracks, and thus their swaths, are spread furthest apart at the equator, and are compressed at the poles. As a consequence, there is an overlap, called *side lap*, of neighbouring swaths. Logically, this side lap is smallest at the equator and increased at the poles.

A basic understanding of ground tracks, swath widths, and side lap is helpful in designing a remote sensing experiment for a few reasons. First, weather and other temporary conditions may prevent good data acquisition for the first and/or "best" pass of a satellite directly over a given location. In this case, side lap may allow for multiple acquisitions of the same location on the surface with only a short delay, provided that the target location can be seen during two or more subsequent passes of the same satellite. This situation is more likely to occur if the sensing target is at high or low latitude than at the equator. Also, side lap may allow experimenters to obtain data for a single location in much more rapid time series than would be possible if the experimenters were to wait for the satellite to exactly repeat its path-a matter of hours, rather than weeks. Finally, side lap may allow the same location to be viewed from slightly different angles at slightly different times of the same day with neighbouring orbital paths. This may provide additional information from shading caused by the sun's angle, and other factors that change with relatively small differences in position and time.

GENERAL IDEA ABOUT ELECTROMAGNETIC SPECTRUM

THE ELECTROMAGNETIC SPECTRUM

The idea of the electromagnetic spectrum is one of those great scientific syntheses of observable natural phenomena that should be part of the knowledge of any scientifically educated person. Other similar syntheses are things like the kinetic-molecular theory of matter, the concept of the vastness of the Universe and our relation to it, the idea of deep time and the evolutionary development of life, the DNA code of life and the ideas of basic biochemistry.

Since the information technologist makes use of the electromagnetic spectrum a great deal, it is appropriate to devote a lecture to it in our course.

Electricity and Magnetism

These were two disparate ideas that were first discovered in antiquity: the Greeks gave the name (as in so many areas) through their term for amber. Amber is a naturally occurring fossizilized tree sap, made famous in the movie Jurassic Park. Anyway, the Greeks had observed that amber when rubbed could attract small pieces of dust or other light trash. Also, static electricity sparks were also known.

Naturally occurring magnetic material ("lodestone" or a type of iron oxide with a higher concentration of iron than usual) was also known to the ancients. It is said that the Chinese were aware that a piece of lodestone could be floated on a piece of cork, and would stabilize to a consistent North-South orientation, thus making possible a compass.

Hewever, it wasn't until the 1700s that the beginnings of our modern "scientific" approach to electricity and magnetism began.

Alessandro Volta noted a connection between electricity and life when he made dissected frogs legs twitch by touching them with dissimilar metals. [This inspired Mary Shelley, the wife of the famed poet Percy Byssche Shelley, to write a famous novel, whose theme was "There are things that Man is not meant to know..."] Anyway, Volta had been experimenting with placing dissimilar metals, *e.g.* copper and zinc, into salt solutions, and obtaining an electric current.

It was early shown that if wire was wrapped into a coil, and connected to a battery, that a magnet was produced in the area close to the coil. If a piece of iron was placed inside the coil, the field was greatly magnified, giving a device called an electromagnet. If the iron was a hard steel alloy, it would retain the magnetism after the current was removed, allowing the manufacture of permanent magnets. Lodestones were quite weak compared to artificial magnets.

Hans Christian Oersted observed that if a wire was wrapped around a compass, the compass would deflect.

This was a unique phenomenon. The uniqueness came from the fact that the interaction was at right angles. The compass needle didn't line itself up with the wire, but at right angles to it. This was the first "vector cross product" relationship discovered. All other interactions hitherto known acted in a straight line between centers: gravitational attraction, electric force and magnetism.

Scientists had been experimenting with pure electric force also. To observe pure electric force out in the open requires charges of thousands of volts, which, fortunately, are easily obtained by friction. When you give someone a shock by "static electricity" you are using thousands of volts. It is not lethal because the amount of electric current is small.

[I use the term "scientists" here. In those days, in the 18th and 19th centuries, the usual term was "philosophers" or "natural philosophers". Our "science" was their "natural philosophy."]

[One can make water analogies of these ideas; in fact any electrical device can have a water analogy. One can think of electric charge as being a quantity of water, current as the rate of flow of water, and voltage as the force or pressure propelling the water. The force or pressure comes from a source of energy; either the water must be pumped up to a height to give it pressure-that is, the water has potential energy, or else a pump must take energy from some other source to force the water to move directly. Our city water system uses a combination of both methods. Strictly speaking, voltage is called potential difference, and represents the potential energy per unit charge, measured between two points in space.]

These studies were being carried out mainly in Europe, especially France and Italy. England was mainly preoccupied with the developing industrial revolution, and its scientific workers were clarifying ideas about energy and thermodynamics, motivated by the development of the steam engine.

America was regarded by European intellectuals as a land of wild Indians and stagecoaches, if they thought about it at all. The idea that someone in America could do Natural Philosophy seemed absurd. However, Benjamin Franklin in the middle of the 1700s had a brief period in his life when he had made enough money from his printing shop to semi retire, but hadn't become embroiled in the American Revolution as representative in France yet.

He experimented with electricity, leading up to his famous kite experiment in which he showed that lightning was the same basic thing as static electricity created in the lab. Since he knew that lab static dissipated quickly if the charged object had sharp points on it, he thought that maybe lightning could be dissipated before the actual stroke if sharp objects electrically connected to the earth were to be placed on high roofs. This was shown to be effective. Some feel that this vivid demonstration of the ability of man to evidently interfere with the "will of God" successfully was a significant force leading to the growth of secularism in Christendom. There was, after all, no more vivid demonstration of the "will of God" than to be struck down by a lightning bolt!

A similar formula was shown to work for idealized magnetic poles:

The constant mu zero is used to make the numbers work out. M1 and M2 are the magnetic pole strengths. [In fact this equation is an idealization more so than the electric equation above. An isolated magnetic pole is an unachievable abstraction-in practice magnets always exist as dipoles-N and S poles fairly close together. Electric poles can in principle be separated as much as needed.]

What all of this is leading up to...

By the middle 1800s quite a bit of electrical technology had been worked out. Things like telegraph lines were beginning to appear, allowing instant communication over vast distances. In fact, the mid 19th century saw communication finally surpass that under the Roman Empire 2000 years earlier!

Electric batteries, motors, generators existed. Electric power stations were starting to appear.

James Clerk Maxwell studied all the disparate physics and engineering involved in this field, and formulated a theoretical formulation that tied it all togther. This led to the great triumph of "field theory" in which each point in space could be thought of as posessing a magnetic and electric force; that is, there existed an electric and a magnetic field. Oersted had shown that these were related, and the electric motor and generator were the practical embodiment of this discovery. Physics students learn about these when they study electromagnetism in 2nd or 3rd year. A good part of the early technical education of a would-be physicist is to learn the necessary mathematical tools, that is, integral and differential calculus, to enable manipulation of these equations. A physicist has to learn how to apply these to specific real-world situations (albeit a bit idealized for pencil and paper calculations). This theory was a revolution in the 1870s when it appeared; it took a generation for it to be absorbed into the technical consciousness of physicists.

When undergraduate physics students first start to work with these equations, there is an understandable "rush" in their minds; all these amazing things revealed by such a small number of symbols. To me, an interesting issue is, Where does the information actually lie? After all, it takes an educational process spanning many years to enable the physicist to work with these equations; is the information in the symbols, or are the symbols a bit like very high level function calls in an API environment, in which they trigger a complex series of mental events that yield the result. The function call without its substrate of lower level operations would be useless. Maxwell's equations without the technical education are just so many hieroglyphics, as they may well seem to CPSC students!

Enter the Electromagnetic Spectrum

Maxwell initially formulated his theory based on an idea of the electromagnetic effect being due to a sort of fluid flow. From the basic equations, he worked out the speed of the interaction; this is a calculation including the two basic physical constants: the electric constant and the magnetic constant. Quantitatively, the velocity comes out to 3 E8 m/s, or the speed of light. This immediately led to the idea that some sort of "electromagnetic" waves could exist, and the possibility that light might be such a wave. Attempts to detect such waves were carried out by Heinrich Hertz, who demonstrated the first apparatus for creating and detecting low frequency electromagnetic waves, *i.e.* "radio".

In 1902, Guglielmo Marconi demonstrated the transmission of radio waves across the Atlantic. This was the beginning of the radio industry. One of the main applications was the implementation of ship to shore telegraph. Previously, a ship was completely out of touch with the world while on its voyage. The sinking of the Titanic in 1912 was an early example of the use of this technology to summon help.

Gradually, engineers learned how better to control radio waves, allowing for "wave bands" or fixed frequency "bands" allowing simultaneous use of electromagnetic waves by different people in the same area for different transmissions. Today, of course, we have this with our radio stations; we "tune them in" by adjusting our receiver to the desired band.[As a side issue, what about the idea of a fluid for electromagnetism? This idea persisted for some decades, and was finally put to rest by the Michelson-Morley experiment, showing that no physical medium existed (or if it did, it carefully moved so as to nullify any attempt to measure the movement of any apparatus through it!) This led to Einstein's Special Theory of Relativity, and the equivalence of mass and energy, and the atomic bomb.]

Frequency Bands

Early radio technology was of necessity very simple minded, one might say crude. Although high frequencies could carry more information the simple electronic valves (*i.e.* "tubes") could only operate at relatively low frequencies, in the megahertz region. Thus, the first broadcast channels appeared around 1 MHz.

A channel can be thought of as an independent path for information. Modulation is the process of coding information into a channel. A technique called "amplitude modulation" or AM allowed audio information to be coded into a radio frequency channel of narrow bandwidth, in fact only double the frequency range of the audio itself. Since audio goes to 20KHz, this would imply that an AM channel of 40 KHz width would be needed. In fact, it was decided to use a bandwidth of only 10 KHz, since this was adequate for the sound reproduction technology back in the 1920s when radio appeared. [Radio killed off much live theater, vaudeville, and the domestic sale of pianos!] Thus, we had about 100 possible channels in the range 600-1600KHz. A problem with all electromagnetic channels is that in any one reception area, only one channel at any one frequency is usable. Because AM radio stations "carry" quite a long way-the lower frequency ones diffract around the earth's curvature, the higher frequency ones bounce off the ionosphere, especially at night, the 100 channels must be spread around the country fairly thinly. If two channels on the same band are picked up at the same time by a receiver, a squealing noise is heard.

Above the AM band, a number of miscellaneous AM bands developed for various services, collectively called “short wave”. The ionospheric bounce was particularly effective in some of these bands, making them useful for military transmissions, and propaganda broadcasts, especially in the “Cold War” from 1950 to 1990. I can remember as a teenager playing with shortwave tuning into Russian rantings (in English) about the “dictatorship of the Proletariat”. I can’t imagine who would want to listen to this.

Relatively recently, a band at about 27MHz was allocated for AM Citizen’s Band radio. This originally had 24 channels, later expanded to 40. You can see how narrow the CB band looks on the diagram, compared to AM radio. This is due to the logarithmic plot I have used for frequency in the horizontal axis. More modern technology has allowed these higher frequencies, where there is more room for channels, to be used economically, with the aid of “crystal filters” and “frequency synthesis” techniques, not available in the 1920s.

An inventor named E H Armstrong invented many of the basic analog circuits used for radio receivers. A problem with AM radio was its lack of noise rejection. He conceived of the use of another technique, frequency modulation or FM. This would need about 10 times the bandwidth per channel, requring the use of higher frequency bands where there would be enough Hz available. He worked with RCA, who gave him the use of a laboratory and a transmitter setup on the roof of one of the new New York skyscrapers then being built. However, as mentioned in a previous page, Sarnoff, the head of RCA, decided to throw his resources into the development of television. Thus, FM lost the support of RCA, Armstrong committed suicide, and TV ended up with much of the spectrum space. You can see FM squeezed in between the TV channels. RCA threw Armstrong a bone by making the TV audio frequency modulated, but the picture is AM.

Each FM band gets 100KHz, which Armstrong felt was a sell-out; he wanted more. Each TV band takes about 4500Khz, with channels about 6000Khz apart. This separation allowed cheap TV receivers to reliably receive adjacent channels in the same market area.

The long carry of short wave stops working at around 20MHz, so that TV and FM stations are basically “line of sight.” This necessitates high transmitting antennas, and allows each major market to have lots of available channels.

In effect, a copy of the electromagnetic spectrum can be carried inside a cable, as an electrical analog, up to frequencies of about 500MHz or so. So, we have cable TV and other services, compatible with the airborne electromagnetic channels. Finally, at 800 MHz, we find the original cell phone band, used for analog phones. The next image shows the two digital cell phone bands. I need to find out what bands WI-FI uses, and the upcoming ultra high speed wireless network bands. This latter, I understand, uses "spread spectrum" technology; the information is carried using many different frequencies rather than on single frequencies. This makes interference a serious problem, except that the power is kept very low, reducing range.

Historically, the TV and Cell bands were originally used only for RADAR (Radio Direction and Ranging). This was developed as a crash programme during World War 2. Like many technologies, it was invented by the British but developed elsewhere, in this case the USA. The British invented the "cavity magnetron" vacuum tube, in which high voltage electrons are injected into a very strong magnetic field, causing them to swirl in a curved path, oscillating at the desired gigahertz frequencies. [When this was first demonstrated to the Americans, one of the leading scientists, I I Rabi, said "It's easy-an analog of a whistle" prompting one of the others there to remark sourly "OK, Rabi, how does a whistle work?"] In fact many audio devices and microwave devices are analogous, since the wavelengths are similar. Treble frequencies for audio have mechanical waves similar in length to gigahertz electromagnetic microwaves.

Radar was very crude in its use of the frequencies, basically just an on-off setup. Our modern cell phones are far more sophisticated in their modulation techniques, allowing thousands of channels in any local area.

Apart from crude things like microwave ovens and heat lamps, a huge stretch of the electromagnetic spectrum is currently unused. One reason is that our atmosphere is opaque to these frequencies. This in fact is a cause of the greenhouse effect. The short wave infrared can penetrate the atmosphere, coming from the Sun. It is absorbed by the earth, and re-emitted as long wave infrared. This is because the earth is much cooler than the Sun, and the wavelength produced by an emitter of electromagnetic radiation shortens as its temperature rises.

We generate radio by electrical excitation. Heat will also generate it, although not in a useful fashion, as there is a mess of random frequencies produced. Frequencies like our TV would be characteristic of temperatures very close to absolute zero.

Finally, we reach the visible band, a very narrow strip from about 400 to 800 THz. This band exists because our atmosphere is transparent in this narrow range, and our eyes thus work in this area. Fortunately, the air blocks ultraviolet except the very near UV. This is fortunate because this is the start of the bands that can disrupt chemical bonds.

We are starting to make use of light in CPSC, for fibre optics. Certainly, there are a lot of Hz there. However, our modulation is still rather crude, although it is improving; in fact, unexpected improvements may have contributed to the great internet "bust" in 2001. Companies laid huge amounts of long distance fibre, expecting that each fibre could carry maybe 5 channels. So, methods of sending 200 channels appear, resulting in a vast oversupply of fibre bandwidth! Channels can be sent over fibre using different colors. This is called "wave division multiplexing". I suppose our AM and TV channels could rejoice in the same term!

Visible light in many ways is a "watershed" in terms of wavelength and frequency. The visible band shows both particle and wave characteristics, depending on what experiment is done to observe. Visible light shows wave effects like diffraction and interference, but also shows particle like effects (*e.g.* the photoelectric effect; Einstein got his Nobel from his explanation of this effect, not from relativity).

A whole treatise could be written on the physical causes of visible colors. Basically, they result from energy transitions within atoms, of electrons moving from one state to another. Colors towards violet are caused by higher energy transitions. As we move into the UV, the transitions can involve molecular disruption. The Ozone Layer blocks near UV; if it gets through the degree of molecular disruption on the surfaces of living things becomes an issue, with skin cancer, etc, possibly becoming a serious problem.

Continuing to higher frequencies, X rays are characteristic of very high energy electron transitions. They are produced in CRTs by the electrons slamming with 20000 Volt potential into the inner face of the tube. The tubes are made with lead glass to block these X rays. "Harder" X-rays come from low level nuclear disruption; the nucleus of the atom is about 6 orders of magnitude smaller than the atom itself, and about 6-7 orders of magnitude harder to change in state. Nuclear changes of state (*e.g.* atomic or hydrogen bombs, or the core of the sun) release gamma rays.

There are isolated extremely high energy gamma rays, about 10 orders of magnitude beyond my diagram, occasionally seen. Their source has been a mystery for decades; I believe some tentative explanations now exist.

Shannon's Law and Data Transfer

The Bell Telephone System ran the telephone network in much of the USA until the late 1970s or early 1980s as a government regulated monopoly-almost a public utility. Since they had a protected position, with a guaranteed profit, they could engage in long term research into issues related to communications.

Since by law they could not sell ancillary discoveries for profit, they basically gave away many of their great developments, such as the transistor. [We might note that their Unix developed just as the Bell system was being broken up by the US Government, and the legal restrictions were also being removed. So, Bell started trying to sell Unix licenses, for an amount about equal to the gross national product of Luxembourg, with the predictable result that several non Unix Unices developed, thus "forking" the development of the system, producing several incompatible versions, and thus aiding Microsoft in its march to hegemony.]

1948 was a great year. The transistor was announced to the world, the Manchester University (England) "Baby" electronic computer ran the first stored programme in June, and Claude Shannon announced his communication theory.

Shannon worked much of his professional life at Bell Labs. His main claim to fame is his study into the error-free information capacity of an analog channel. (We mustn't forget as we think about digital technology that it all is really analog underneath). Analog operation involves continuous functions of variables such as voltage, current, time.

Noise is any unwanted energy; it arises from many causes. The most basic is thermal noise caused by the random motion of atoms and molecules in all matter. Noise comes in from the cosmos also; some of the "snow" visible on a TV with an antenna if tuned to a vacant channel may be from this source [Some of this may be echoes of the Big Bang!]. Not surprisingly, human technology produces lots of noise; any electrical switching or sparking produces electromagnetic noise. Computers are of course changing state billions of times per second; each state change is physically done by voltages within the system changing. This will propagate electromagnetic noise. [In fact, there are regulations regarding the amount of RFI (Radio Frequency Interference, or noise) that can be produced by any electronic device. One problem with all this sort of regulation is that compliance is expensive; a small company must send its device to a consultant lab which will run the necessary tests and sign the forms that go to the regulators, a process that may cost $10,000 to $100,000. This process is especially onerous in Europe; where some have cynically suggested that one reason for the stiff regulations is that the big industries, like Siemens and Philips, can maintain staffs for dealing with compliance issues, while small companies face a serious barrier to entry in the marketplace as a result. In North America, the regulations are not so stiff, and in Asia, it is still the "Wild West" in many places.]

Signal to noise ratio is the ratio of the signal value (voltage, current, energy, pressure, or whatever) divided by that of the noise. I will use a value of 1000 in the examples below. This corresponds roughly to AM radio, or the telephone on a good day. High fidelity audio amplifiers can achieve values of around 100,000, limited by thermal noise in the transistors and resistors. [Professionals often measure this in decibels. For voltage or current measurements, the decibel (dB) value can be determined by multiplying the log (10) of the S/N by 20, so a S/N of 1000 would be 20x3 or about 60dB, and 100,000 would be about 100dB.]

Finally, let's consider light. The visible spectrum is about 4 x 10 ^ 14 Hz wide. If we use the same 1000 S/N we get a gross data capacity of 4 x 10 ^ 15 bits/sec, or about 4 x 10 ^ 14 bytes/sec, or about 4 x 10 ^ 8 Megabytes per second, or about 4 x 10 ^ 5 gigabytes per second, or about 4 x 10 ^ 4 DVD movies per second!

Unfortunately, this assumes that we can control light the way we can control radio frequencies, with detailed control over each electromagetic "vibration". This is not possible with light due to things like "Doppler broadening" caused by the thermal motion of the atome producting the light. However, much progress can be made from our current technology.

ELECTROMAGNETIC ENERGY

Remotely sensed data are collected in many regions of the electromagnetic spectrum. Data recorded from each part of the spectrum can provide distinct information on characteristics of the Earth's surface or properties of the atmosphere. For example, healthy green vegetation reflects highly in the near-infrared region of the spectrum, whereas water bodies tend to reflect only a small amount of incoming radiation in the visible region. All remote sensing instruments collect electromagnetic radiation that is reflected, emitted, or scattered from the Earth's surface and atmosphere. So-called active sensors such as radar and lidar emit energy that bounces off the land or water surface and returns to the sensor to be recorded. The way the energy is directed or scattered by the surface, and the time it takes for the energy to return, reveals information about surface characteristics. Because of the long wave lengths employed by radar, the signals can penetrate clouds, thereby allowing scientists to record information about normally cloud-covered areas. This is an asset in tropical areas such as the Amazon River basin.

Passive sensors, on the other hand, typically rely on solar illumination of the Earth's surface, though some are equipped to detect night-time lights and gas flares. These sensors are "passive" because they do not emit their own energy, but rather rely on energy reflected or emitted from the earth's surface. Unlike radar sensors, they are unable to penetrate clouds. It is interesting to note that the visible portion of the spectrum-those wavelengths that humans can see-is a very small segment of the spectrum. Part of the strength of remote sensing is that it enables scientists to "see" portions of the spectrum that are outside the range that the human eye can detect. Scientists can combine non-visible portions with visible ones through colour composites, assigning each *band* (or portion of the spectrum detected by the instrument) the colours red, green and blue.

A sensor's bandwidth and the number and placement of bands (within the spectrum) define its *spectral properties*. Panchromatic sensors measure reflected energy in a single portion of the electromagnetic spectrum, usually the visible to near-infrared regions. Multispectral sensors, on the other hand, collect reflectance information in discrete portions of the spectrum, with each being recorded as a separate image called a band or channel. When these bands are displayed on a computer, with one band shown through each of the blue, green and red channels of the monitor, they yield a combined colour image. Landsat

7's Enhanced Thematic Mapper, for example, is a multi-spectral instrument that collects data in eight bands-three visible (one each for blue, green, and red), a near-infrared, two middle-infrared bands, a thermal-infrared and a higher spatial resolution panchromatic band. By contrast, the Moderate Imaging Spectrometer (MODIS) collects data in 36 different spectral regions, and the Airborne Visible/Infrared Imaging Spectrometer (AVIRIS) is a hyperspectral instrument that collects data in 224 spectral bands.

Ultimately, what a sensor measures is the intensity of radiation that actually reaches the sensor, which is termed the *at-satellite radiance*. Radiance values are commonly translated into digital numbers (DNs). The possible range of DNs varies between sensors, although ranges of 0-255 (for 8 bit images) and 0-1023 (for 10 bit images) are common, with higher values corresponding to greater brightness. Radiance is captured by a two dimensional array of picture elements, or *pixels*. A DN for a pixel in a specific band is determined by the intensity of the radiance captured for that particular portion of the electromagnetic spectrum. If space-based passive sensors were able to accurately, precisely and repeatedly capture the actual reflectance from a feature on the ground, regardless of the time of day, season or weather conditions, much of the hard work of image processing would be eliminated. But the reality is that the atmosphere scatters radiation that is reflected back out to space. Smoke, haze, clouds and humidity exacerbate the problem, and can block reflected energy entirely. Data from shorter wavelengths are more likely to be blocked or scattered by clouds or atmospheric particles, whereas images using sensors capturing longer wavelengths are less likely to be disturbed by atmospheric conditions between the sensor and the target object.

THE PROMISE OF REMOTE SENSING

Remote sensing advocates argue that the flaws inherent in data sets traditionally used in land cover analysis can be corrected by using satellite images to measure the "real" cover of the ground. These traditional data, which include census data and modelled estimates, are indeed problematic and their unclear genealogies and biases make them analytically questionable. For example, many allegedly authoritative assessments of global forest and grassland cover depend upon data reported by the World Resource Institute, the Global 2000 Report, and World Watch research papers, which are, in turn, based upon national self-reporting and state census (World Resources Institute 1992). These are commonly biased through the interests of the state to reflect rhetorical and propagandistic coverage requirements; where forests are politically useful, for example, they often appear in data sets. These data are also collected under dissimilar methods, which makes comparison difficult. Moreover, these data are assembled in categories that are inherited relics of historical environmental management that reflect arbitrary conveniences in administration. Understanding land cover dynamics is difficult if one is forced to rely on these kinds of statistics.

Efforts in geographic information science and remote sensing, on the other hand, according to its advocates, can "develop methodologies for global land cover classifications that are objective [and] reproducible...". Poor sampling and biased reports will, in this way, be rectified through a controlled analysis of remotely sensed data to create accurate category sets.

The resulting images, freed from the bias of politics, will appear to provide "a message without a code" and knowledge of the Earth's surface free from the prejudice of authorship. By ground-truthing, assessing for accuracy, and employing recently developed post-classification corrections, the resulting classification supposedly becomes a good match between the image and the land covers the analyst is looking for on the ground.

Although such remote sensing methods show promise, associated landscape classification faces technical obstacles. Estimates of land cover change based solely on satellite data, for example, are limited in the richness of vertical information, especially in thickly layered and complex ecologies.

Similarly, mismatches in spatial and temporal scale and resolution make estimates of change difficult to assess and compare. Even so, remote sensing promises comprehensive inventory and analysis, free of the subjective filters of society and politics.

CLASSIFICATION AS METAPHOR

But the categories of remote sensing analysis are no freer from inherited meaning than those of state-level aggregated statistics. Rigorous image processing can mathematically cluster pixels with like reflectance values but must ultimately rely on an assignment of meaning to such clusters that is inter-textual; it has been inherited from other information and knowledge sources and is historically arbitrary.

The arbitrary character of these land cover categories is rooted in their role as metaphors for complex biotic and non-biotic landscape features, which give common meaning to disparate objects. In landscapes characterized by heterogeneity, features are gathered into or excluded from common types by deciding that some relevant characteristic for distinction makes them essentially the same or different. Such characteristics ultimately must be drawn from the metaphoric vocabulary of the analyst.

The Inevitability of Metaphors in Classification

Consider a typical false colour composite image (assembled from Green, Red, and Infrared bands) of a set of agricultural villages adjacent to a hilly area. This visualization of a complex data set shows high levels of variability in digital pixel values across three spectral ranges of reflectance. Certain features in the image will be apparent to the trained eye; the red hills may be heavily vegetated, the winding gray paths extending from the hills might suggest dry streambeds, and darker blue polygons might be towns. The act of interpretation has already begun before any formal analysis, clustering, or classification is conducted. The targets of our classification, those things we already consider real, are winnowed in advance: forests, streambeds, and towns.

To assign formal qualitative meaning to what we see in the image still requires a vast number of decisions about what pixels are most like one another, but the metaphors for understanding are already in place. Unsupervised classification, for example, uses a clustering algorithm to form clusters from the undifferentiated mass of pixels, in an apparently objective fashion. The algorithm guarantees that, based on scrutiny of a multi-dimensional histogram, pixels share membership in categories of "like" reflectance.

The number of clusters must be decided in advance, although in most cases, the standard procedure is to overestimate the number of categories, cluster the data, and then aggregate pixel groups together into meaningfully "different" sets. To do so, the analyst surveys the ground, and perceives that "these pixels represent landscapes that are similar, or enough like something I have seen, or seen defined elsewhere; thus, they are effectively the same." Defined by such standards, pixels are lumped together and aggregated to become an authentic image of the Earth's surface.

Multiple unique locations become "forest," for example, in the moment that they are named in the same way. This then, is an act of metaphoric deployment. There is no true forest per se, only multiple discrete things brought together as if they were the same; by conjoining and naming them together, they become the same, however, and are accepted as literal in the moment of mapping.

Supervised classification works in a reverse fashion; places known to be what the analyst recognizes as "forest," for example, are used to create training sites, a cluster of pixels of a known cover type. Reflectance signatures are derived from these sites and undifferentiated pixels are categorized by similarity to the training signature, using a range of mathematical decision rules, such as minimum or maximum values or likelihood based on probability.

Here again, the concept or trope of forest, an ideal type, is used to name, constitute, and bound cartographic entities, formed prior to analysis. Iterative techniques, such as Isodata clustering, neural networks, and tree-based classifiers that partition and "prune" categories, combine elements of supervised and unsupervised classification, going back and forth between composited images and raw data to produce coherent clusters.

Like unsupervised classification, these clusters must still be aggregated and named. Like supervised classification, these methods utilize carefully chosen training sites—areas of already known land cover—to check against resulting classifications.

MODELING METAPHORS: FUZZY SETS AND LAND CHARACTERISTICS DATABASES

Methods do exist that take seriously the ambiguous nature of classification and the problems inherent in grouping the chaotic differences of the landscape. These include fuzzy set classification and "land characterization database" approaches. Fuzzy set classification of remotely sensed data incorporates uncertainty by valuing and displaying pixels based on the probability of membership in a given spectral group using a Beyesian or Dempster-Shafer procedure. In this way, contested membership in metaphoric land classes is reflected as a probability surface. This innovation attempts to incorporate uncertainty and ambiguity into planning and decision making for landscapes in which a unanimous interpretation does not exist. Even so, the technique can only reflect our uncertainty about the membership of any given pixel and location in any predefined set of land cover classes.

Land characteristics databases, as used in a prototype project at EROS data centre, are premised on the idea that land cover may be broken down into discrete, and seasonally distinct, elemental characteristics, including specific species mixes, ground surface cover, and canopy, which can then be combined to suit the needs of the analyst. These characteristics can, furthermore, be later aggregated to link local and regional studies to more global questions and to serve as a "rosetta stone" to translate between varying classification schemes. Nonetheless, while offering a departure from a single, arbitrary, classification technique, the land characteristics method is not free from the problems posed by more traditional techniques. The discrete elements that form the basis of the typology, be they the soil, hydrology, or canopy cover, are themselves identified before the fact.

Classification is ultimately deductive then, rather than inductive; categories known in the mind's eye are found in the data, not vice versa. Even the most empirical clustering technique requires the naming and consolidation of clusters based on judgement or definition.

The data, however carefully differentiated, therefore, do not "speak for themselves," and more sensitive algorithms provide only an increasingly advanced and careful reproduction of existing categories, not the "discovery" of new ones.

In all cases then, the landscape vocabulary of the analyst is a necessary prerequisite to classification, and the analyst relies on an imaginary landscape against which to check resulting differentiation. This landscape is assembled from cover concepts that the analyst must already perceive based on features such as canopy, vegetation mixes, or apparent emptiness. Recognition of this fact is not new in debates over land cover classification. In consolidating and institutionalizing the USGS land use/land cover classification system, Anderson et al. emphasized that "there is no one ideal classification of land use and land cover... the process itself tends to be subjective even when an objective numerical approach is used." Anderson et al. also stressed that in aggregating and collapsing variable covers into a limited number of cover types, the interpreter must use subjective "patterns, tones, textures, and shapes" known before the fact.

Classification Metaphors are Cultural Artifacts

Unacknowledged, however, is the fact that such landscape "flags" and perceptions are culturally bound; as a classic example, an Inuit imagines and perceives the difference between upsik and api snow cover, where other observers might not. The larger conceptual language or "paradigm" of the observer drives the particular simplifications selected in analysis. The culture of forestry and remote sensing, into which a global population of analysts with similar training and background fall, ultimately produce and reproduce a common set of metaphors, and a common set of categories. Such categories reflect unproblematic classes that appear "universal" to the practitioners who share a common culture. These vary surprisingly little from case to case and reflect inherited notions of meaningful land classes. To illustrate, consider the range of categories conventionally deployed in remote sensing and land cover change analysis. The most common cover categories, drawn from a survey of papers published in prominent international remote sensing journals between 1995 and 1999 include: Barren Land, Water, Grasses, Mixed Forest, Open Shrublands, Agricultural, Evergreen Broadleaf Forest, Forest Land, Wooded Grasslands/ shrubs, and Urban or Built Up. These categories appear in a majority of the papers, and while the studies represent a range of geographical and ecological contexts, only a handful of categories dominate the analysis. As Reed et al. (1994) note, "scientists typically select a land cover framework based on availability rather than suitability to a problem" or context. Similarly, the Anderson et al. (1976) classification itself, with its multi-tiered category system, allows a fine grade of distinctions, but its lower-level detailed categories are less often used in analysis. This occurs despite new techniques, such as image segmentation and filtering, which enable finer and more specific category sets.

This kind of uniformity in classification is an inherent and normal process of both daily experience and scientific analysis in a world of chaotic differences. There is nothing unscientific about the assignment of meaning that creates such categories of likeness. As in other arenas of "normal" science, consensus on such categories is necessary to order explanation and is part and prerequisite to analysis. At the same time, however, these associations, which define "which element pertains to which set," are fundamentally cultural and follow differing logic in differing contexts, both within and outside the remote sensing laboratory. Accepting even the most common and self-evident categories like "forest," "grasses," and "barren land" for the study of the distribution and change of landscapes without querying their socio-cultural sources or alternatives is problematic. The danger is not only that our analytical vocabulary is unnecessarily impoverished but, it is, moreover, that these culturally determined cover categories will inevitably reveal certain features, important to some observers, while disguising features that may well be important or relevant for other observers.

Categories as State Control

Competition between category sets is typically resolved through the imposition of a formal set of land cover classes over those of an already existing, in situ, cultural system. In the process, the imposed categories serve as a lever for control and prescription, and often act to reproduce the patterns of degradation and landscape change that rational and scientific planning seeks to halt. This problem occurs, to some degree, in the U.S. and many other developed world contexts; the culturally bound categories of the analyst or expert often contradict and confront those of the residents or communities for which images and maps are being produced. Such conflicts are even more apparent in the colonial and post-colonial context, where foreign categorical systems have been deployed uniformly over existing ones.

Land Cover Control in Colonialism

Colonial mapping entailed implicit (and sometimes explicit) unification of diverse spaces into uniform territory and thus required coherent and consistent land cover categories. The resulting maps were products of scientific state land management which, although itself a product of European science, was initially implemented exclusively in the colonial periphery. In these areas, from as early as the 18th century, conservation forestry began to exert a tremendous influence on colonial states, with a power unknown in Europe. In British colonies, such authority was especially marked. Drawing on the work of Alexander von Humboldt and other geographers, scientific connections between deforestation, drought, and economic crisis justified complex and stringent systems for land cataloguing and management. The goal of scientific forestry was to preserve the resource for public good through stringent techniques of naming and recording.

More than simply applying names, however, land classification under colonialism resulted in a political mapping of control, which linked the formation of "what is"—land capacities and characteristics—to what "should be"—land potentials and optimizations. This "survey modality" fixed and bounded natural objects in terms of how they ought to be used. Through the practice of surveying and mapping, colonial power accumulated bodies of information, created standard land cover types, and established these as norms by which the landscape must be judged and managed. Land covers imposed control by eliminating competing understandings.

4

Remote Sensing Application in Crop Husbandry and Agriculture

REMOTE SENSING APPLICATIONS IN CROP ASSESSMENT AND FORECASTING

The Central Sector Umbrella Scheme *viz.* 'Forecasting and Remote Sensing Applications in Crop Husbandry' during the Tenth Five Year Plan includes following three components.

- National Crop Forecasting Centre (NCFC)
- Crop Acreage and Production Estimates (CAPE)
- Special Data Dissemination Standards (SDDS).

The umbrella scheme of Forecasting and Remote Sensing Application in Crop Husbandry comprising three ongoing components of NCFC, CAPE and SDDS was implemented w.e.f 1.8.2004. Before 1.8.2004, these three components were being operated as individual schemes.

NATIONAL CROP FORECASTING CENTRE (NCFC)

NCFC was created at the end of the year 1998 and was set up mainly with the following functions: Periodic crop forecasting for major crops, and Co-ordination and assimilation of various methodologies and technical advancement relating to crop forecasting.

However, over the time, its sphere of activities has been expended to: Providing effective unified institutional framework for the entire crop forecasting system in the country involving data flow, assimilation, analysis and dissemination of statistics.

Periodic crop forecasting for major crops through assimilation of information generated by the different organizations such as IMD, Medium Range Weather Forecasting of Department of Science and Technology, Department of Space (DOS), Central Statistical Organization (CSO), Indian Agricultural Statistics Research Institute (IASRI), Field Operations Division of the National Sample Survey Organization (NSSO) and the State Agriculture

Statistics Authorities (SASAs), State Agriculture Departments etc. Co-ordination and assimilation of various methodologies and technical advancement relating to crop forecasting. Central level monitoring of the situation about crop, weather, supply of inputs, pests/diseases and related aspects through the mechanism of Crop and Weather Watch Group in the Department of Agriculture and Cooperation.

Providing a forum for the Standing Technical Committee on Agricultural Statistics to review and monitor the development of the methodologies for crop forecasting in particular and Agricultural Statistics in general, which has already been constituted as a follow up of the recommendation No. 8 of the Expert Group. Coordinating the proposed projects/scheme on "Development and Application of Extended Range Forecast System for Climate Risk management in Agriculture (ERFS)" and "Forecasting of Agriculture output using Space, Agro-meteorology and Land based observations (FASAL)".

During the Ninth Plan, the professional posts envisaged for NCFC could not be created and NCFC operated by ad-hoc deployment of professionals with a truncated mandate. This activity was initiated during the 7th Five year Plan as one of the components of the Scheme named 'Remote Sensing Application Mission for Agricultural Application' (RSAMAA).

The Scheme was initially monitored by the Crops Division, but later on, it was transferred to Directorate of Economics and Statistics. From 2004-05, this has been merged with the umbrella scheme Forecasting and Remote Sensing application in Crop Husbandry.

This activity is being fully funded by the Ministry of Agriculture and executed under the overall technical guidance of the Department of Space, Ministry of Science and Technology, Government of India, with the help of State Remote Sensing Application Centres (SRSACs), State Department of Agriculture (SDA), Directorate/Bureau of Economics and Statistics (DESs) and State Agriculture Universities (SAUs). It aims at the estimating crop acreage and yield, with the application of RS Technology, at least a month before the actual harvesting of crops. In this process, it enables development and upgradation of methodologies in consonance with state of art RS technology and sensor capabilities for crop inventory assessment at different geographical units.

Under the CAPE activity, area and production estimates based on remote sensing technology have been prepared for specified crops for the selected States/Districts during the year 2004-05. Grants-in-aid are provided to Space Application Centre, Ahmedabad for operationalising the activities under CAPE.

Special Data Dissemination Standards (SDDS)

Under this activity, quarterly estimates of agricultural production are generated for use in the compilation of Quarterly National Accounts by the

Central Statistical Organization. This activity has been undertaken in order to meet the obligations concerning supply of data to the International Monetary Fund. The estimates of quarterly crop production generated are being furnished to the Central Statistical Organization. In the absence of direct data, quarterly production is estimated by using the estimates of Kharif and Rabi seasons in conjunction with crop calendar. In order to improve upon the quality of quarterly estimates by way of refining the estimation procedure and cross validation of results, available data from other sources such as Timely Reporting Scheme, market Intelligence Unit of DES, National Sample Survey Organization etc..are used. This is a staff oriented activity under the umbrella scheme.

Market Intelligence and Price Analysis

The DES also collects and compiles wholesale and retail prices, international prices and market arrivals of essential commodities on weekly/ monthly basis for 700 centres (for wholesale prices) and 83 centres (for retail prices) spread all over the country. Out of these price quotations, about 216 quotations on wholesale prices are being disseminated through E-mail to the Office of Economic Adviser, Ministry of Industry, for construction of Wholesale Price Index Numbers on weekly basis. The price quotations collected are also disseminated in the form of weekly bulletins and circulated through E-mail to concerned offices. Besides, daily wholesale prices of some important agricultural products, in respect of Hapur and Delhi are also being supplied to All India Radio for Broadcasting.

APY (Area Production and Yield)

The objective of this component is to obtain estimates of area and production of principal crops, in each season, with a break up of area under irrigated/unirrigated and traditional/high yielding varieties of crops on the basis of priority enumeration conducted in a sample of 20 per cent of villages by a specific date. The States are required to furnish these estimates by 30th November for Kharif Crops and by 30th April for Rabi crops. These estimates are used for generating advance estimates of production of principal crops. This component is being implemented in 16 land record States and also Union Territories of Delhi and Pondicherry.

The component has funding pattern on 50:50 basis between the Central Government and the State Governments. The estimates of area and production of principal crops for Kharif and Rabi seasons for the agricultural year 2001-02 have been received from all the implementing States/Union Territories under the component. The performance of the component of the main scheme namely Improvement of Agricultural Statistics is also being closely monitored through Quarterly Progress Reports and Seasonal Progress Reports. The Budget Estimates of 2002-03 have provided ₹ 42500 thousand as the share of the Central Government.

AGRICULTURAL STATISTICS AND AGRICULTURAL POLICY FORMULATION

The budget provided for the scheme during the year 2002-03 is ₹ 17500 thousand. The objective of this scheme is to strengthen the system of agricultural statistics and policy formulation by strengthening the research techniques and upgrading of skills of personnel involved in the compilation and analysis of data. During the year 2002-03, a National Workshop on Improvement of Agricultural Statistics was held on 26-27 June 2002 which was inaugurated by Shri J.N.L. Srivastava, Secretary (Agriculture and Cooperation). The workshop was attended by a large number of representatives from Ministry of Agriculture, Central Statistical Organisation, National Sample Survey Organisation, Department of Space, Commission for Agricultural Costs and Prices, State Governments, Research Institutions like IASRI and State Agricultural Universities. The Workshop discussed various issues including the need to reduce the divergence in successive estimates of crop production, increased application of remote sensing technology in agriculture sector, new methodologies for estimation of fruit and vegetable production and estimation of small area statistics. The workshop also discussed the recommendations of the National Statistical Commission (NSC) on the agriculture statistics, and it was agreed that States would take appropriate action in the matter.

This Directorate has brought out the annual publication "Agricultural Statistics at a Glance" in June 2002. This is a very useful publication and is used as a compendium for data on almost all important aspects of Indian Agriculture by planners, agricultural scientists and research scholars. The scope of the latest issue has been enlarged by including sections on horticulture and plantation crops as well as a few new tables on State-wise poverty line, time series data on annual growth rates in agriculture, commodity-wise and year-wise exports/imports, key economic indicators, etc. The various Sections in the publication have been made more reader friendly. This publication is widely circulated and highly appreciated by users.

CAPE (CROP ACREAGE AND PRODUCTION ESTIMATION)

CAPE is a Central Sector Scheme. It was initiated during the 7th Five Year Plan as one of the components of the scheme named 'Remote Sensing application mission for agricultural application' (RSAMAA). The scheme was initially monitored by the crops division, but later on was transferred to DES. The scheme is funded by the Ministry of Agriculture and operated by the Dept. of Space, Ministry of Science and Technology, Govt. of India. The scheme aims at the application of space Technology for estimating crop acreage and yield at least a month before the actual harvesting of crops. In this process, the scheme enables development and upgradation of methodologies in consonance with state of art RS technology and sensor capabilities for crop inventory assessment at different geographical units. As the advance estimates of area and production of crops are required for taking policy decisions on procurement storage and pricing measures, remotely sensed data has an immense potential in the timely monitoring of the crop acreage and production at district/group of districts/ regional levels due to its wide area synoptic coverage using the Remotely sensed data at the peak vegetative stage of crops, it is possible to give pre harvest estimates of crop acreage at the right time.

COC (COST OF CULTIVATION)

The Comprehensive Scheme for studying the Cost of Cultivation of Principal Crops in India is being implemented since 1970-71 as a 100 per cent Central Sector Plan Scheme on the basis of recommendations made by the Standing Technical Committee on Indices of Input Costs constituted in 1967 under the Chairmanship of Dr. Ashok Mitra. The main objectives of the scheme are as follows:-

- Collection and compilation of field data on cost of cultivation and cost of production in respect of 29 crops.
- Estimation of crop-wise and state-wise costs of cultivation and production.
- Computation of the indices of terms of trade between agriculture and Non-agriculture sectors.

The estimates of cost of Cultivation are used by Commission for Agricultural Costs and Prices (CACP) for recommending Minimum Support Prices (MSPs) of 25 crops. Directorate of Economics and Statistics (DES) in the Ministry of Agriculture gets this study conducted through 16 Agricultural/ General Universities/Colleges besides the Directorate of Tobacco Development. Under this scheme, the field data pertaining to the cost of cultivation/ cost of production are collected, compiled and tabulated by different Implementing Agencies. The data are checked, corrected, validated and analysed to generate crop-wise and state-wise cost estimates by DES before these are passed on to the CACP to be used as data base for recommending the MSPs of various agricultural commodities to the Government of India.

The cost estimates generated under the scheme are also used for policy formulation by the Central Government and State Governments, and others including Agricultural/General Universities, Government/ Non-Government Research Organizations, and individual researchers. In addition, the data are also used for working out the Index Numbers of Terms of Trade between Agriculture and Non-agricultural Sectors.

The scheme is being implemented in 16 states, namely, Andhra Pradesh, Assam, Bihar, Gujarat, Haryana, Himachal Pradesh, Karnataka, Kerala, Madhya Pradesh, Maharashtra, Orissa, Punjab, Rajasthan, Tamil Nadu, Uttar Pradesh and West Bengal. The scheme covers 29 crops,namely, paddy, wheat, jowar, bajra, maize, ragi, barley, moong, urd, arhar, gram, masur, groundnut, rapeseed/ mustard, nigerseed, soyabean, sunflower, sesamum, cotton, jute, sugarcane, VFC tobacco, onion, potato, ginger, tapioca, areacanut, coconut and black pepper.

The field data under the scheme are collected on the Cost Accounting Method under which daily entries of debit/credit for the expenditure/income are made in order to assess the total cost incurred/benefit accrued by/ to each farmer covered under the scheme. The field data are collected by the field-men. The detailed questionnaires are filled up/updated on monthly/annual basis after making enquiries on a daily basis from 10 farm holdings which consist of 2 each from 5 different size classes *i.e.* upto 1 hectare, 1-2 hectares, 2-4 hectares, 4-6 hectares and above 6 hectares allotted to each field-man.

During the 9th Five Year Plan an expenditure of ₹.687700 thousand was incurred under the scheme. The actual expenditure during 2000-01 and 2001-02 was ₹.138500 thousand and ₹.189500 thousand respectively. An allocation of ₹ 985000 thousand has been proposed for the 10th Five Year Plan. Out of total allocation of ₹ 185800 thousand for the year 2002-03, an amount of ₹ 77600 thousand has already been released by September, 2002 as the first installment of grant-in-aid. The scheme is proposed to be implemented on block grant basis from 2003-04.

No specific activities are undertaken for the women and for the North-Eastern States under this Scheme as it is meant for collection and compilation of data on cost of cultivation and cost of production of agricultural crops. In the North- Eastern Region, the scheme is in operation in Assam only through the Assam Agricultural University, Jorhat for which an amount of ₹ 5800 thousand has already been released during the year 2002-03 out of an allocation of ₹ 11500 thousand. The proportional contribution of principal crops to the national total production by other North-Eastern States is not significant enough for conducting study in these States.

CES (CROP ESTIMATION SURVEY)

This component before merger with Umbrella scheme was a Centrally Sponsored Plan Scheme of the 9th Five Year Plan. This is being implemented since the Sixth Plan to generate estimates of area and production of fruits, vegetables and minor crops in the country. The component is presently being implemented in 11 States and a total of 14 crops are covered: 7 each of fruits and vegetables. The methodology developed by the Indian Agriculture Statistical Research Institute, is being used for generation of these estimates. Budget Estimates of ₹ 40000 thousand has been provided for the year 2002-03 under this component.

The different fruits and vegetables which are being covered are: apple, mango, citrus, pineapple, grape, banana, guava, potato, onion, cabbage, cauliflower, tomato, ginger and turmeric. The component is being implemented in the States of Andhra Pradesh, Gujarat, Haryana, Himachal Pradesh, Karnataka, Maharashtra, Orissa, Punjab, Rajasthan, Tamil Nadu and Uttar Pradesh.

SPAC (SUPPORT PRICING FOR AGRICULTURAL COMMODITIES)

The main objective of the Government's price policy for agricultural produce is to ensure remunerative prices to the growers for their produce with a view to encouraging higher investment and production and to safeguard the interests of consumers by making available supplies at reasonable prices. The price policy seeks to evolve a balanced and integrated price structure in the perspective of the overall needs of the economy. To achieve this end, the Government announces, each season, minimum support prices for major agricultural commodities and organizes purchase operations through public and cooperative agencies such as Food Corporation of India (Paddy, Wheat and Coarse Cereals), Jute Corporation of India (Jute). Cotton Corporation of India (Cotton), National Agricultural Cooperative Marketing Federation of India (NAFED) (Pulses and Oilseeds) and Tobacco Board (Tobacco) besides other agencies designated by the State Governments.

The Government decides support prices for various agricultural commodities after taking into account the recommendations of the Commission for Agricultural Costs and Prices, views of State Governments and Central Ministries as well as such other relevant factors, which in the opinion of the Government are important for fixation of support prices.

The Government has already announced Minimum Support Prices for Kharif Crops of 2002-03 Season. Keeping in view the recommendations of Commission for Agricultural Costs and Prices (CACP), the Minimum Support Prices (MSPs) of paddy, coarse grains (Jowar, Bajra, Maize and Ragi), Arhar, Cotton, Copra and Soyabean have been retained at last year's level.

The MSPs of Moong, Urad, Sunflowerseed have been raised by ₹.10 per quintal. The MSP of Groundnut-in-shell has been increased by ₹.15 per quintal.

As compared to previous season, the MSPs of Nigerseed and Sesamum have been raised by ₹.20 and ₹.50 per quintal respectively. The MSP of Jute has been raised by ₹.40 per quintal as compared to last year. In addition, the Government has announced Special Drought Relief Prices (SDRP) for kharif marketing season 2002-03 based on the recommendations of CACP which took in to account the losses suffered by farmers in un-irrigated areas and additional expenditure incurred by the farmers in irrigated areas. The SDRPs announced by the Government are intended to benefit farmers in drought hit areas. These range between ₹.5 and ₹.20 per quintal for major agricultural commodities. These prices do not form a part of MSP and would be by way of relief and not by way of any bonus. The price policy pursued by the Government has paid rich dividends by way of increased production, investment and also helped in generating exportable surpluses in respect of certain agricultural commodities.

The MSPs fixed for major agricultural commodities in recent years alongwith Special Drought Relief Prices for Kharif Crops of 2002-03 season are shown in the enclosed statement.

The price policy for Rabi crops of 2002-03 to be marketed in 2003-04 season is, at present, under consideration of the Government.

REMOTE SENSING AND ITS IMPORTANCE IN AGRICULTURAL SURVEY

Remote sensing is nothing but a means to get the reliable information about an object without being in physical contact with the object. It is on the observation of an object by a device separated from it by some distance utilizing the characteristics response of different objects to emissions in the electromagnetic energy is measured in a number of spectral bands for the purpose of identification of the object.

In such study single tabular form of data or map data is not sufficient enough which can provide can be, combined with information's obtained from existing maps and tabular data.

- Remote Sensing techniques using various plate form has provide its utility in agricultural survey
- Satellite data provides the actual synoptic view of large are at a time, which is not possible from conventional survey methods.
- The process of data acquisition and analysis is very fast through Geographic Information System (GIS) as compared to conventional methods.

Remote Sensing techniques have a unique capability of recording data in visible as well as invisible (*i.e.* ultraviolet, reflected infrared, thermal infrared and microwave etc.) part of electromagnetic spectrum. Therefore certain phenomenon, which cannot be seen by human eye, can be observed through remote sensing techniques *i.e.* the trees, which are affected by disease, or insect attack can be detected by remote sensing techniques much before human eyes see them.

PRESENT SYSTEM OF GENERATING AGRICULTURAL DATA AND ITS PROBLEMS

The present system of agricultural data is collected throughout the nation. The main responsibility of collection agricultural survey lies on the Director of Land Records, Director of agriculture and District Statistical Office under the Ministry of Agriculture. These data are collected not only on a local but also some extent of district and state level. The associate of agricultural survey on crops (crop production, type of crop and crop yield), range land (condition of range, forest type, water quality, types of irrigation system and soil characteristics) and livestock (livestock population, sex of animal, types of farm and distribution of animals).

The basic problems in this survey are;

- Reliability of data
- Cost and benefits
- Timeless
- Incomplete sample frame and sample size
- Methods of selection
- Measurement of area
- Non sampling errors
- Gap in geographical coverage
- Non- availability of statistics at disaggregated level.

Remote Sensing techniques make it use before the remote sensing data may provide solution to these particular problems of agricultural survey.

REMOTE SENSING TECHNIQUES FOR AGRICULTURAL SURVEY

The given factors influenced the use of remote sensing in agricultural surveys; via 1. Characteristics of the agricultural landscape 2. Characteristic of EMR on Agricultural survey. Detection, identification, measurement and monitoring of agricultural phenomena are predicated on the assumption that agricultural landscape features (*e.g.* crops, livestock, crop infestations and soil anomalies) have consistently identifiable signatures on the type of remote sensing data.

Some of the parameters which may cause these identifiable signatures include crop type, state of maturity, crop density, crop geometry, crop vigour, crop moisture, crop temperature, soil moisture, soil temperature. An image analysis can correlate a certain signature with one of these many characteristics. Remote Sensing techniques in agriculture survey which affect the signature on remote sensing imagery. It is important to consider briefly the significance of choosing the appropriate sensor system, as well as the scale and resolution requirements that will yield optimum benefits for objectives of agricultural survey.

ADVANTAGES OFREMOTE SENSING TECHNIQUES IN AGRICULTURAL SURVEY

With the primary aim of improving the present means of generating agricultural data, a number of specific advantages may result form the use of remote sensing techniques. Vantage point. Because the agricultural landscape depends upon the sun as a source of energy, it is exposed to the aerial view and, consequently, is ideally suited or remote sensing techniques. Coverage. With the use of high-altitude sensor platforms, it is now possible to record extensive areas on a single image. The advent of high-flying aircraft and satellites, single high quality images covering thousand of square miles

Permanent record. After an image is obtained, it serves as a permanent record of a landscape at a point in time which agriculture changes can be monitored and evaluated.

Mapping Base. Certain types of remote sensing imagery are, in essence, pictorial maps of the landscape and after rectification (if needed), allow for precise measurement (such as field acreages) to be made on the imagery, obviating time-consuming on the ground surveys.

These images may also aid ground data sampling by serving as a base map for location agriculture features while in the field, and also as a base for the selection of ground sampling point or areas.

Cost savings. The costs are relatively small when compared with the benefits, which can be obtained form interpretation of satellite imagery.

Real-time capability. The rapidly with which imagery can be obtained and interpreted may help to eliminate the lock of timeliness which plagues, so many agricultural survey. Other advantages of Remote Sensing:

- Easy data acquisition over inaccessible area.
- Data acquisition at different scales and resolutions
- The images are analysed in the laboratory, thus reducing the amount of fieldwork.
- Colour composites can be produced from three individual band images, which provide better details of the area then a single band image or aerial photograph.
- Stereo-satellite data may be used for three-dimensional studies. At present, all advantages listed above have been demonstrated either operationally or experimentally:

APPLICATION OF REMOTE SENSING TECHNIQUES FOR AGRICULTURAL SURVEY

The specific application of remote sensing techniques can be used for i) detection ii) identification iii) measurement iv) monitoring of agricultural phenomena.

Area of Specific Applications

- Applicable to crop survey
  - Crop identification
  - Crop acreage
  - Crop vigour

- Crop density
- Crop maturity
- Growth rates
- Yield forecasting
- Actual yield
- Soil fertility
- Effects of fertilizes
- Soil toxicity
- Soil moisture
- Water quality
- Irrigation requirement
- Insect infestations
- Disease infestations
- Water availability
- Location of canals.

• Applicable to range survey:
- Delineation of forest types
- Condition of range
- Carrying capacity
- Forage
- Time of seasonal change
- Location of water
- Water quality
- Soil fertility
- Soil moisture
- Insect infestations
- Wildlife inventory.

• Applicable to livestock survey:
- Cattle population
- Sheep population
- Pig population
- Poultry Population
- Age sex distribution
- Distribution of animals
- Animal behaviour
- Disease identification
- Types of farm buildings.

The use of remote sensing technology has been rapidly expanded for the development of key sectors. This chapter highlights the fact that the remote sensing techniques will continue to be very important factor in the improvement of present system of acquiring agricultural data. The remote sensing provides various platforms for agricultural survey. Satellite imagery has unique ability to provide the actual synoptic vies of large area at a time, which is not possible for conventional survey methods and also the process of data acquisition and analysis are very fast through GIS (Geographic Information System) as compared to the conventional methods. The different features of agriculture are acquired by characteristic, spectral reflectance, spectral signature of agriculture and associated phenomena through EMR. In General the research paper emphasizes the utmost need of timeliness and accuracy of the output generated by remote sensing techniques and its calibration with ground-truth and other information systems like aerial photography and satellite imagery etc. Further, the importance of remote sensing with special reference to agricultural sector involving crop acreage, crop production, rangeland and livestock.

SPACE INPUTS FOR PRECISION AGRICULTURE

Precision agriculture, newly emerging agricultural management concept, embodies the convergence of biotechnologies and other agricultural technologies with space and informatics. With a goal to achieve the quantum jump in agricultural productivity, reduced cost of cultivation, diversified and resilient agricultural systems, the precision agriculture plays catalytic role in order to achieve a common ground between environmental and economic goals. It is basically designed to optimize agricultural inputs *viz.*, fertilizers, pesticides, water etc, in tune with micro-level/field requirements. Optimization is focused on increased yields, reduced cost of cultivation and to minimized environmental impacts through location-specific management.

The success stories pertaining to Precision Agriculture have mainly drawn from the developed countries; wherein Agriculture is characterized by highly mechanized and automated systems, and is driven by market forces and has been professionally managed enterprise.

Taking into account the predominance of fragmented land holdings, heterogeneity of crops and livestock and concept of farm families in the rural conditions, the model of Precision Agriculture representing the typical Indian Agricultural scenario is yet to evolve. While the ecological integrity of farming systems is an imperative need, it is equally important to extend the access of information and market to the small and marginalized farmers. The Precision Agriculture model for India while addressing these issues should provides an innovative route for sustainable agriculture in globalize and liberalized economy.

TECHNOLOGY ELEMENTS

Recent advances in technology for variable rate technology (VRT), with concurrent advances in remote sensing, GIS & GPS, and the developments taken place in crop simulation modelling, have provided enough opportunities and scope to take-up proto-type Precision Agriculture experiments. The VRT applies production inputs at rates appropriate to soil and crop conditions within micro-level field conditions. The VRT systems have been demonstrated for several materials, including herbicides, fertilizers, insecticides and seeds.

Role of space technologies becomes more crucial in order to address the spatial variability of soils and crops across the various scales of mapping. The space technology inputs also capture the vulnerability and dynamism of agricultural systems.

The developments in space-borne imaging sensors, particularly their spatial, spectral and temporal resolutions are well characterized to capture these features. While high spatial resolution images enable mapping and monitoring the structural attributes of agro-ecosystems, high spectral resolution or hyper-spectral imaging addresses their functionalities. The high temporal resolution captures the dynamisms of agro-ecosystems.

The use of remote sensing, GIS and GPS for crop monitoring, condition assessment and yield modelling has already been well established. Crop simulation models (CSMs) provide potential production under the different scenario of constraints, including weather, soils, crops, cultural practices etc. The conjunctive use of VTR, remote sensing, GIS, GPS and CSMs provides technological framework for Precision Agriculture.

SPACE APPLICATIONS TO AGRICULTURE

Agriculture has been at the top of our priorities for space applications. The road map of agricultural applications started with the first remote sensing experiment on identification of coconut root-wilt diseases in Kerala using Infrared aerial photography-way back in 1969-70. Since then, space applications to agriculture sector have touched almost all the segments of agricultural ecosystems. These include the mapping and monitoring of major crops, soils/ degraded lands, command areas, wastelands, surface and ground water, floods and drought, and watersheds.

Agricultural statistics that provides the vital informatics base to agriculture sector originates from age-old village level Patwari system. It often moves upward with its inherent subjectivity and bias. There are hundreds of crucial decisions at different levels, which are taken purely based on this. In order to strengthen the foundations of agricultural statistics in the country, a remote sensing based the nationwide mission called pre-harvest Crop Acreage and Production Estimation (CAPE) was launched in late 80s.

Covering all the major cereals, pulses and oilseeds, CAPE provides in-season crop statistics with 90/90 accuracy at state level. The CAPE could thus provide the scientific basis of agricultural statistics and is transitioning further to yet another institutional destination entitled Forecasting Agricultural Output using Space Agro-meteorology and Land-based Observations (FASAL) within Ministry of Agriculture itself.

Synthesizing the state-of-the-art in econometrics, agricultural meteorology and remote sensing based modelling, FASAL envisages the multiple productions forecasting of the major crops with improved accuracy to the extent of 95/95 criteria, timeliness and scope in terms of covering the whole country. The FASAL could be used to provide scientific solutions facilitating crop insurance, bridging the gaps between crop production and post-harvest technology, pricing and policy decisions.

Extensions of irrigation, genetically improved crops and use of inorganic fertilizers have been the harbingers of Green Revolution in the country. Those multi-purpose irrigation commands, which accelerated the irrigation networks in 70s and 80s, are now characterized by depleted irrigation efficiency, water logging and salinity.

At the behest of Ministry of Water Resources, remote sensing based command areas inventory has been taken up in the priority irrigation commands to enhance the water use efficiency. In the similar line, a coarse scale land degradation mapping was carried out at behest of Ministry of Agriculture. It is important to mention here that in some of critically affected areas of Gangetic plains of Uttar Pradesh, land degradation to the extent of village level has been mapped out. Based on this, several land reclamation efforts-including the World Bank supported programmes, have been taken up to restore the health of soils. In coming years, with the support from Ministry of Agriculture, ISRO/DOS is planning to launch Nationwide Land Degradation Mapping Mission at the scale 1: 50,000, so as to delineate land degradation at Village level.

By virtue of the unique combinations of IRS satellites with varied resolutions and capabilities, the space segment in the country is in a position to monitor agricultural drought and recurrent floods more efficiently. Capturing the flood events with the damages to the Villages and crops has formed the scientific basis for damage assessment and thereby relief operations at various levels.

This year, for example, flood inundation maps in Orissa were generated depicting the marooned areas during the different flood waves and disseminated to the user community within hours through the Internet. Space inputs to agricultural drought monitoring is yet another aspect to target the districts and taluks on the basis of severity of drought.

Department of Space at the behest of Ministry of Rural Development has carried out nation-wide wasteland mapping on 1: 50,000 scale using the IRS data. The Wasteland Atlas of India covering the wasteland statistics of entire country has been brought out. These maps would help in retrieving the information at village/watershed (500 ha) levels, for implementation of wastelands/watershed development programmes. The uniqueness of wasteland mapping has also been the creation of digital database with village as well as watershed boundaries.

With the hydro-geomorphological maps prepared under the National Drinking Water Mission, and subsequently to the Rajiv Gandhi National Drinking Water Mission, satellite remote sensing has really made a dent in reaching its potentials down the line to grassroots. Search for groundwater, particularly in areas with consolidated and semi-consolidated rock formations, considered more difficult from the point of view of exploration as well as recharging of groundwater, is considerably aided by the use of the hydro-geomorphological maps. These maps are extensively used for locating prospective groundwater sites around problem habitations in the country, as a part of the 'scientific-source finding'.

At the behest of the Planning Commission, scientific inputs were generated from remote sensing on land use/land cover of entire country to evolve agro-climatic zonation based planning. These applications have moved further to large scale mapping across the problem regions of the country (especially the rain-fed regions), towards micro-level decision making for sustainable development. Under remote sensing based Integrated Mission for Sustainable Development Mission (IMSD), land and water resources development plans were carried out for about 80 mha. in 175 districts of the country.

Implementation of these plans in certain areas has enriched the ground water potential, increased the cropping intensity along with the net returns from the fields. It is important to note that these experiments were conducted in the terrain of low productivity, mainly dry lands/rain-fed areas, predominantly wastelands, where the scope of doubling food production is the main issue. Emphasis is being laid in operational utilization of this vast database for ground level implementation by concerned user agencies.

The success of these experiments lies not only due to the applications of space inputs alone, but largely due to the synergistic interfaces among grass root level farmers, administrators/policy makers and multi-disciplinary scientific communities. With operationalisation of the National Natural Resources Management System (NNRMS) in the country, with DOS as the nodal agency, the space, ground and user segments have been properly tuned to respond to the challenges of sustainable crop production in the country.

The INSAT-VHRR observations on clouds, cyclone depressions, surface radiance and monsoon parameters in the Indian Ocean have really strengthened the agro-meteorological services in the country. To disseminate the appropriate locale-specific agricultural packages, there are several programmes such as 'Krishi Darshan' being beamed through all the regional channels.

An operational system of public instruction has been established since 1995, using the INSAT based Training and Development Communication Channel (TDCC) for disseminating the improved agricultural practices, training primary school teachers, Panchayat Raj, elected representatives, Anganwadi workers, wasteland development functionaries etc. The Jhabua Development

Communication Project (JDCP), undertaken by ISRO in Jhabua district of MP, has been providing communication support to the developmental activities and also interactive training to the development functionaries, besides empowering the poor and marginalised tribal.

A pilot project sponsored by the Planning Commission, a single window information service provider concept-called Agro-climatic Planning and Information Bank (APIB) has been developed for the farmers of drought prone areas of Karnataka.

The APIB is an effort to empower the farmers and reduce their vulnerability to crop failures. Moving to the farmer's cooperatives, ISRO has taken up jointly with IFFCO-a major initiative on developing remote sensing and GIS based decision support system for fertilizer movements in the priority areas. Finally, space applications over last three decades in agriculture have made this sector information rich and have addressed the various issues encompassing the different sectors of agriculture.

INDIAN PRECISION AGRICULTURE

Agriculture in India, as we see today, is at the crossroads.

- On the one hand there are depletions of ecological foundations of the agro-ecosystems, as reflected in terms of increasing land degradation, depletion of water resources and rising trends of floods, drought and crop pests and diseases. On the other hand, there is imperative socio-economic need to have enhanced productivity per units of land, water and time.
- At present, 3 ha of rain fed areas produce cereal grain equivalent to that produced in 1 ha. of irrigated. Out of 142 million ha. Net sown areas, 92 million ha. are under rain-fed agriculture in the county.
- From equity point of view, even the record agricultural production of more than 200 Mt is unable to address food security issue. A close to 60 Mt food grains in the storehouses of Food Corporation of India (FCI) is beyond the affordability and access to the poor and marginalized in many pockets of the country.
- Globally, there are challenges arising from the Globalization especially the impact of WTO regime on small and marginalized farmers.
- Some other unforeseen challenges could be anticipated global warming scenario and its possible impact on diverse agro-ecosystems in terms of alterations in traditional crop belts, micro-level perturbations in hydrologic cycle and more uncertain crop-weather interactions etc.

At this stage, agriculture needs new paradigms to deal with the present situation. The strategy lies in integration of the dynamic information and scientific knowledge into the management of agro-ecosystems, and thereby optimizing the radiation, water and nutrient usages. Agriculture has to transition from high inputs material inputs to the optimum level, through the appropriate use of information, knowledge and strategies for efficient resource usages. In such case, productivity of agricultural may not be the function of the quantum of agricultural input use alone, but will include information, knowledge and efficiency while managing the agricultural practices.

With the fragmentations of land holdings and predominance of small and marginal farmers, our agricultural systems are basically dis-aggregated farm families. Fundamentally, Precision Agriculture aims at a dis-aggregated micro-level farm management strategy with intense information inputs addressing the variations of soils, crops, water, chemicals, market access etc.

Taking into the present state of Agriculture in the country, Precision Agriculture is absolutely essential in order to address poverty alleviation and food security to a very large cross-section of the population. For example, at present, the average Orissa farmer produces slightly more than one ton of rice per hectare, and keeps little below one tone for his family. But if he can produce four tons, then he has three tons to sell, and more cash in hand. The smaller the farm, the greater the need for a marketable surplus. Defining Precision Agriculture in Indian Context Taking into account the typical characteristics of Indian Agriculture, the definition of Precision Agriculture, therefore, must encompass the strategy and framework to achieve higher productivity, reduced cost of cultivation by optimization of inputs, and diversified and resilient agricultural systems.

All these goals are to be achieved within the typical constraints of India's agro-ecosystems. While the depletion of ecological foundations of farming systems needs to be arrested, the access of information, credit, agricultural inputs and market to the small and marginal farmers are equally crucial. The Precision Agriculture model for must be derive encompassing all these issues, within a broad framework of addressing the negatives of globalization and achieving sustainable agriculture in the long run.

The definition of Precision Agriculture has to be micro-level contextual and should capture the local variabilities, vulnerabilities and dynamisms of agro-ecosystems. For example, Agriculture in Punjab, Haryana and Western UP is characterized by higher productivity (about 4 t/ha), higher use of inputs (irrigation ~ 96 per cent; fertilizer consumption 0.158 t/ha), higher cropping intensities and predominance of medium and large farmers.

Sustainability of such agro-ecosystems has been a cause of concern in the recent times.

The model for Precision Agriculture for such ecosystems may focus on sustainability through the optimization of agricultural inputs and thereby reduction of cost of cultivation, and cropping system analysis.

Agriculture in Southern part of the country-especially Andhra and Tamilnadu is moderate yield (around 2 t/ha), inputs (irrigation ~ 55 per cent;

fertilizer consumption 0.12 t/ha) and cropping system based, with predominance of small and marginal farmers. Precision Agriculture for such system should aim at enhancing the productivity based on in situ soil and water conservation.

The Eastern India especially Orissa, though high potential agricultural system is unfortunately characterized by low inputs (irrigation ~ 25 per cent, fertilizer use 0.025 t/ha), low yield (~1.2t/ha) and low cropping intensity, with predominance of marginal and small farmers.

A model for Precision Agriculture for Orissa could focus more on poverty alleviation and food security through enhanced productivity and low cost of cultivation. The models for Precision Agriculture in Indian context could therefore be the micro-level and contextual-addressing the empirical issues in the diverse agro-ecosystems of the country. The access and outreach of best practices in Precision Agriculture-as demonstrated by contextual models, to the large cross-sections of the farming community would then help in cultivating the new paradigms in age-old agriculture enterprise in the country.

Precision Agriculture models are not complete, unless the parameters related to empowerment of the farmers; especially small and marginal farmers are integrated. In this context, ISRO has also initiated Gramsat project in Orissa. In the line of JDCP, the Gramsat project aims at empowering the people especially the poor and marginalized, by awareness building and access to information and services. Towards this, a network of one-way video and two-way audio Village Information Kiosks is being developed in the selected blocks of Orissa. The same networks are also planned to facilitate e-governance in the region. Precision Agriculture model should present a synergy that could lead to a holistic mission, focused on agricultural development with the backdrop of present issues and challenges.

5

Rainwater Harvesting for Agricultural Water Supply

TECHNICAL DESCRIPTION

This is the most common and one of the earliest methods of rainwater harvesting uses. In this method, an earthen bund with a height and width of 30 to 45 cm is constructed around a field. Rainwater fills the *aman* created within the bund; an *aman* is a monsoonal paddy grown in June/July and harvested in October/November.

The monsoon season begins in June/July, or some times in August if the monsoon comes late.

With water stored on the field for few days, the soil softens and is suitable for preparation using ploughs drawn by animal power or power tiller. If the field retains excess water, it may be flushed out by cutting a portion of the bund.

Once the land has been prepared, aman seedlings are transplanted or seeds are broadcast spread. In hilly areas, this technology can be applied in valleys between hills or wherever there is level ground on which water may be stored for few days before planting.

Extent of Use

All farmers whose lands are fed by harvested rainwater during the monsoons make use of this technology for preparing their lands. Farmers, by experience, have learned to match the aman crop cycle with the rainfall pattern, and the cropping cycle may be delayed by a month or so if monsoon season starts late in any particular year. This method not only softens the soil but also assists in controlling weeds by inundating organic matter and exposing weeds, crop residuals and other organic substances to aquatic decomposition processes. The organic matter is then mixed into the soil, thereby supplementing soil nutrients. Nearly all types of farms, large and small, use this process. The rainwater satisfies about 78 per cent to 97 per cent of the water requirements for land preparation for aman crops; the other farms, about 3 per cent and 22 per cent of farms, that are unable to match their cultivation with the onset of the monsoon rain, bring water from ponds or khals, rivers, wells or other waterbodies to prepare the fields for aman crops.

In saline areas, harvested rainwater is used to provide agricultural water for aman paddy cropping until the river water becomes sweet. In these areas, harvested rainwater meets around 50 per cent of water requirement for the aman crop, with the balance of the water being obtained from river sources.

Operation and Maintenance

The bunds need to be repaired every year.

Level of Involvement

This technology is implemented at the household level and requires little external involvement.

Costs

One time construction costs incurred in the creation of the bunds around an acre of land amount to six days of hired labour, or, at the 1996 wage rate of $1.50/day during harvesting period, about $9.00. Locally available agricultural equipment can be used for the bund construction.

Fig. Excess Water is Discharged Through the Bund.

Annual repair costs typically involve two day of family labour per year at an equivalent cost of about $2.50. Costs per acre may be calculated at between four and five days of family labour and from three to six days of hired labour, or about $7 per season on average.

Effectiveness of the Technology

The information obtained during the field survey is suggests that paddy yields decline in cases where rainwater was not used to prepare the agricultural lands for planting. Declines in yields ranged from 44.8 per cent to 51.4 per cent of the yields obtained in fields that were prepared using this technology. To achieve similar yield as obtained with this technology, farmers would have had to incur additional expense.

Suitability

This technology is suitable for use in all flat areas with adequate rainfall.

Advantages

The technology uses locally available materials and is inexpensive to implement.

Disadvantages

The principle weakness of this technology is that it is dependent on rainfall, which may be uncertain. This may delay crop planting with consequential reductions in yield. Where there is no outlet for draining excess water from the bunded field, flooding may damage or destroy the standing crop.

Cultural Acceptability

Being a traditional technology, rainwater harvesting for agricultural supply is well accepted.

RAINWATER HARVESTING FOR COMMUNITY WATER SUPPLY

TECHNICAL DESCRIPTION

Rainwater is collected from rooftops of buildings using corrugated galvanized iron (CGI) sheets as the roofing material, a half-cut HDPE pipe gutter, and HDPE down-pipe to collect rainwater in ferrocement storage tanks. Some tanks have separate tapstands.

This technology was introduced into Nepal for the first time in 1988. As a pilot project, a 20 m^3 ferrocement tank was built to collect water from the roof of a middle school in the Village of Daungha.

The system helped to fulfil the drinking water demand of about 300 students and teachers. The success of this community-based system, encouraged the villagers to build more such systems, including an additional storage tank for the middle school. Two further 20 m^3 storage tanks were constructed to harvest rainwater from the Village Committee Office building and Primary School building

Extent of Use

This technology is an example of a community-based rainwater harvesting system in Nepal. The example is from the Village of Daungha, in the Gulmi

District in West Nepal. The interest of villagers and their active participation have helped to promote the development of such rainwater harvesting systems in Nepal, and there are currently 11 ferrocement tanks for rainwater storage in the region.

Operation and Maintenance

The following maintenance work is required in the operation of this technology:

- Regular cleaning of the rooftops and gutters: Rainwater is generally considered to be free of contamination, and, hence, it requires no treatment. Nevertheless, the roofs and gutters should be cleaned regularly to remove particulates and accumulated materials. Rainwater of the first few hours of the beginning of rainy season should not be collected in the tank, but should be used for flushing the roofs and gutters. Asbestos cement sheets and metal sheet roofing coated with lead-based paints should be avoided as they may be dangerous to health.
- Frequent cleaning of the storage tanks: The storage tank is the most expensive part of any rain water catchment system and determining the most appropriate capacity for any given locality and catchment area will critically affect both its cost and the amount of water available for supply. To ensure the longevity of the tank and quality of the water supply, regular cleaning of the tank should be carried out.
- Inspection of gutters and feeder pipes and valve chambers to detect and repair leaks: Rooftops are commonly used as catchments, even though ground catchments can provide a larger catchment areas, yield a greater volume of water to be collected, and are cheaper to construct. However, the use of ground catchments competes with agriculture for available land.

  Hence, in Nepal, both flat roofs, with tiles or plastered concrete leading to a floor drain, and sloped roofs are used. Sloped roofs are preferred to flat roofs because sloped roofs are accessible only for cleaning and repair purposes, and the harvested rainwater has less chance of being contaminated. Regular inspection of the drainage systems and conveyance systems minimizes water loss.

However, because this technology can be constructed and maintained locally, the Nepalese projects have been handed over to the users committees which bear the overall responsibilities for operation and maintenance of the systems.

Such works as may be beyond the capacities and means of the users committees are carried out by the District Water Supply Offices.

Level of Involvement

Implementation of this technology has involved both the local communities and the government.

Costs

Based upon the experience in Nepal, the total cost of a 20 m^3 water supply project, supplying 1 224 l/day, is about \$24 620, or about \$121 per person. Although the operation and maintenance costs of the system are negligible, the capital cost is too high for individual households and a rural community to invest independently in such a system. Thus, the cost of these community-based systems was divided between the government and community in the following ratio:

Government	- \$22 560
Village	- \$ 2 060.

The cost of a 20 m^3 ferrocement tank is about \$2 000.

Effectiveness of the Technology

A total of 183 m^3 of rainwater is collected annually in the storage tanks served by 160 m^2 of rooftop catchment area in the Daungha Village. Assuming that the water is used for drinking purposes only, it is estimated that about 73 m^3 of water could meet the needs of the 34 residents for a year. With this system, water collected during the monsoon helps meet demand during the dry season.

Of the remaining volume of water harvested, the per capita allocation of six litres per capita per day is considered to be too low to completely meet the demand for water for personal hygiene. However, the estimated 20 litres per capita per day supply for domestic use could be met from a household level 22 m^3 rainfall harvesting system. Such a system would require a rooftop area of 49 m^2 roof area and would provide water to an average household of seven members.

Suitability

This technology is suitable for use in areas with adequate rainfall and in villages having a cluster of roofs or large buildings.

Advantages

The main advantage of having a community-based rainwater harvesting system is the time savings accrued in fetching water. This frees up time which may be utilized for other economic activities. Women, who traditionally gather water, gain more time for child care, social activities and income generating activities.

Disadvantages

The disadvantage of this technology is its high initial cost and per capita cost in small settlements.

Further Development of the Technology

While the technology may be considered to be fully developed, in order to determine the potential rainwater supply for a given catchment, reliable rainfall data (mean annual rainfall and its distribution) are required for a period of at least 20 years. Improved rainfall distribution analysis methods could enhance the utilization of this technology.

RAINWATER HARVESTING FOR IRRIGATION WATER SUPPLY

TECHNICAL DESCRIPTION

The two types of agricultural water storages are observed. One type is used to store water on plots which are slightly inclined; water accumulates in the low lying portions of the fields or in a low lying areas nearby.

Water stored using this method is used for both land preparation for aman paddy cropping and during the growing/milking stage of paddy farming when the water requirement is most critical. *Aman* is transplanted in June/July and harvested in October/November.

The other type of water storage involves the storage of rainwater in ditches or depressions located outside bunds or in land depressions beside road embankments. Water is carried from these depression to the fields in buckets, pitchers or traditional dhoon whenever it is needed.

In some cases, if the waterbody is large enough, water may be used to irrigate a rabi crop or for fishing during the period of monsoonal inundation. In some upland areas, rainwater is stored in low lying plots between two hills for use in times of necessary.

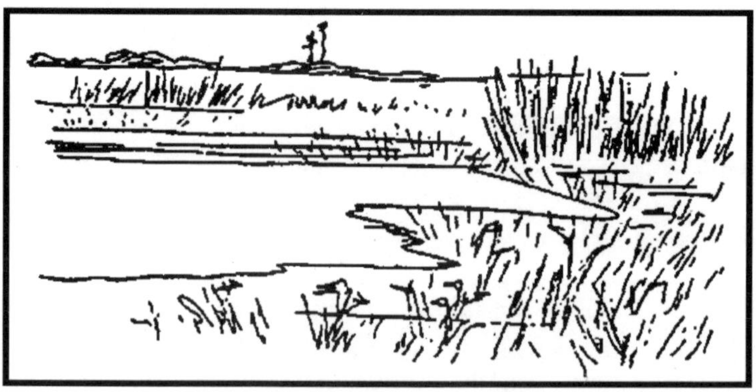

Fig. Stagnant Water on the Lower Slope of a Plot.

Rainwater stored using this technology may be transferred between fields by overflowing successive plots and may be collected in the lowest fields in a given area. In the West-Cental region, rainwater is harvested from lands situated at higher elevations and conveyed to storage ponds through culverts.

In this region, because farm lands are situated at lower elevations than the storage ponds, stored rainwater may also be used for irrigating farm fields during the dry season.

The specific location of the catchment area and its elevation relative to the fields to be irrigated are important factors in determining the potential for using these types of rainwater harvesting technologies.

Fig. Irrigation of Crops with Rainwater in Upland/Hilly Areas..

Extent of Use

Rainwater storage in low lying portions of farms or in neighbouring plots is possible only when the locations and elevations of the plots are suited to such storage.

No detailed data on the extent of use of this technology were identified, but the field survey suggested that, if such opportunities exist, farmers made use of these technologies. In upland areas, nearly all farms having access to rainwater stored in depressions within hills made use of such water for irrigating aman paddies.

Fig. Irrigation of Crops with Rainwater Using Traditional Methods.

In saline areas, this practice was observed on lands located within polders or embankments erected to obstruct the intrusion of saline water. In these areas, there is a conflict between aquacultural (saline water-based shrimp culture) and agricultural (freshwater-based crop) production, which has led to violent confrontations. During the field visit, the murder of a local womens' leader, named Korunamoey Sardar and aged around 40 years, allegedly by shrimp growers has caused continued deep resentment among the local farmers and has strengthened the resolve of the local people to continue fresh water cultivation within the polder in defiance of the threats by shrimp cultivators.

Operation and Maintenance

The use of natural depressions does not eliminate the maintenance requirements. Repair work, such as repairing bunds, maintaining slopes, etc. involve about two person-days of family labour per year at an equivalent cost of $2.50, based upon a rate of $1.25 per day. Local agricultural equipment is used for this purpose.

Level of Involvement

This technology is implemented at the household level, although it has been implemented sometimes at the community level.

Costs

Construction costs are minimal since this system of rainwater harvesting is practised only when the land is naturally sloped, when the farmer has a plot nearby with a low enough elevation, or when there is an existing ditch or depression available where rainwater can be stored.

Effectiveness of the Technology

Data from the North Central zone shows that the use of this technology increases paddy crop yields by about 32 per cent. In order to achieve similar yields as can be achieved through irrigation using rainwater harvesting techniques, farmers would have to incur additional expenses in transporting water from other sources.

The cost of transporting water would be the equivalent of five to six person-days of family labour and one to two days of hired labour.

Suitability

This technology is suitable in areas where natural depressions are found and where there is adequate rainfall to produce run-off.

Advantages

The advantages of this technology are that it is a locally implemented technology, which can be used at little additional cost.

Disadvantages

The disadvantage of this technology is that rainwater cannot be stored for long time due to seepage losses. The method also lacks scientific rigour.

Cultural Acceptability

This technology is very well accepted and has no known cultural disadvantages.

Further Development of the Technology

This is a fully developed technology. However, development of criteria for sizing storage areas would make this technique more rigourous.

ARTIFICIAL RECHARGE OF GROUNDWATER

TECHNICAL DESCRIPTION

Artificial recharge is the planned, human activity of augmenting the amount of groundwater available through works designed to increase the natural replenishment or percolation of surface waters into the groundwater aquifers, resulting in a corresponding increase in the amount of groundwater available for abstraction. Although the primary objective of this technology is to preserve or enhance groundwater resources, artificial recharge has been used for many other beneficial purposes.

Some of these purposes include conservation or disposal of floodwaters, control of saltwater intrusion, storage of water to reduce pumping and piping costs, temporary regulation of groundwater abstraction, and water quality improvement by removal of suspended solids by filtration through the ground or by dilution by mixing with naturally-occurring groundwaters.

Artificial recharge also has application in wastewater disposal, waste treatment, secondary oil recovery, prevention of land subsidence, storage of freshwater within saline aquifers, crop development, and streamflow augmentation.

A variety of methods have been developed and applied to artificially recharge groundwater reservoirs in various parts of the world.

The methods may be generally classified in the following four categories:

- Direct Surface Recharge Technique.
- Direct Subsurface Recharge Technique.
- Combination surface-subsurface methods, including subsurface drainage (collectors with wells), basins with pits, shafts, and wells.
- Indirect Recharge Techniques.

Direct surface recharge techniques are among the simplest and most widely applied methods. In this method, water moves from the land surface to the aquifer by means of percolation through the soil. Most of the existing large scale artificial recharge schemes in western countries make use of this technique which typically employs infiltration basins to enhance the natural percolation of water into the subsurface.

Field studies of spreading techniques have shown that, of the many factors governing the amount of water that will enter the aquifer, the area of recharge and length of time that water is in contact with soil are the most important. In general, these methods have relatively low construction costs and are easy to operate and maintain.

Direct subsurface recharge techniques convey water directly into an aquifer. In all the methods of subsurface recharge, the quality of the recharged water is of primary concern. Recharged water enters the aquifer without the filtration and oxidation that occurs when water percolates naturally through the unsaturated zone.

Direct subsurface recharge methods access deeper aquifers and require less land than the direct surface recharge methods, but are more expensive to construct and maintain. Recharge wells, commonly called injection wells, are generally used to replenish groundwater when aquifers are deep and separated from the land surface by materials of low permeability.

All the subsurface methods are susceptible to clogging by suspended solids, biological activity or chemical impurities.

Recharge wells have been used to dispose of treated industrial wastewaters, to add freshwater to coastal aquifers experiencing saltwater intrusion, and to force water under pressure into permeable bedrock aquifers to arrest land subsidence resulting from extensive withdrawals of groundwater, although with variable success.

In many places, including the United States, Japan and Thailand, the use of injection wells is still considered experimental.

Combinations of several direct surface and subsurface techniques can be used in conjunction with one another to meet specific recharge needs.

Indirect methods of artificial recharge include the installation of groundwater pumping facilities or infiltration galleries near hydraulically-connected surface waterbodies (such as streams or lakes) to lower groundwater levels and induce infiltration elsewhere in the drainage basin, and modification of aquifers or construction of new aquifers to enhance or create groundwater reserves.

The effectiveness of the former, induced recharge method depends upon the number and proximity of surface waterbodies, the hydraulic conductivity (or transmissivity) of the aquifer, the area and permeability of the streambed or lake bottom, and the hydraulic gradient created by pumping.

Using the latter technique, aquifers can be modified by structures that impede groundwater outflow or that create additional storage capacity.

Groundwater barriers or dams have been built within river beds in many places, including India, to obstruct and detain groundwater flows so as to sustain the storage capacity of the aquifer and meet water demands during periods of greatest need.

Construction of complete small-scale aquifers also seems feasible. Notwithstanding, indirect methods generally provide less control over the quantity and quality of the water than do the direct methods.

EXTENT OF USE

The concept of artificial recharge has been known for a long time. The practice began in Europe during the early nineteenth century. However, the practice has rarely been adopted on a large scale, with most large scale applications being found in countries such as the Netherlands, Germany, and USA.

Israel transports 300 million cubic metres of water annually from north to south through the National Water Carrier System and stores two-thirds of it underground. The water is used to meet high summer demands and offers a reliable source of supply during dry years. On the North Plain of China, which is prone to droughts, water from nearby streams is diverted into underground storage areas with capacities of about 500 million cubic metres.

Several counties in Hebei Province are using artificially recharged aquifers to combat sinking water tables. In India, subsurface storage has caught on as a way of providing a reliable source of irrigation water. A number of artificial recharge projects have been carried out in that country.

Operation and Maintenance

To ensure the effective and efficient operation of an artificial recharge system, a thorough and detailed hydrogeological study must be conducted before selecting the site and method of recharge.

In particular, the following basic factors should be considered: the locations of geologic and hydraulic boundaries; the transmissivity, depth to the aquifer and lithology, storage capacity, porosity, hydraulic conductivity, and natural inflow and outflow of water to the aquifer; the availability of land, surrounding land use and topography; the quality and quantity of water to be recharged; the economic and legal aspects governing recharge; and the level of public acceptance.

Level of Involvement

Because of the technical complexity involved in siting and regulating artificial recharge, this technology is generally implemented at the governmental level.

Costs

Rushton and Phadtare describe artificial recharge pilot projects in both alluvial and limestone aquifers in Mehsana area of Gujarat, India. Recharge was accomplished using spreading channels, percolation tanks and injection wells. Table presents a summary of the initial and operational costs for the various schemes. The most expensive scheme, an injection well feeding an alluvial aquifer, had initial and operating costs per unit volume of recharged water of $100/m^3.

Table. Costs of Various Artificial Recharge Schemes in India ($/m³).

Artificial Recharge Structure	Initial Cost	Running Cost
Injection well (alluvial area)	100	100
Spreading Channel (alluvial area)	9	10
Percolation Tank (alluvial area)	2	7
Injection well (limestone area)	6	21
Spreading Channel (limestone area)	7	6

It is apparent from Table that injection wells in hard rock areas are less expensive since they tend to be shallower and have a lesser risks of clogging. Percolation tanks appeared to be least expensive in terms of initial construction costs; this would be the case in areas where the tanks already exist. In such cases, the initial cost only involves the cleaning of the bed of the tank.

For economic reasons, the main uses of artificially recharged water are likely to be providing water for domestic needs, industry and environmental conservation. Because of its relatively high cost, recharged water is not generally suited for irrigation for a total crop, but it can be used to provide supplemental irrigation water for rain-fed crops or to provide additional water to crops at a crucial growth stage during periods of water shortage.

As a general rule in this regard, groundwater must be efficiently used and effectively applied such that the net benefits from its use are maximized over time. Guidelines for the socio-economic and financial appraisal of artificial recharge projects in developing countries, necessary to assess these net benefits, are provided by CGWB.

Suitability

Groundwater recharge methods are suitable for use in areas where aquifers exist. Typically, unconfined aquifers are recharged by surface injection methods, whereas confined aquifers are generally recharged through subsurface injection. Surface injection methods require relatively flat or gently sloping lands, while topography has little effect on subsurface recharge methods. Aquifers best suited for artificial recharge are those which can absorb and retain large quantities of water. In temperate humid climates, the alluvial areas which are best suited to artificial recharge are areas of ancient alluvium, the buried fossil river-beds and interlinked alluvial fans of their main valley and tributaries.

In the arid zone, recent river alluvium may be more favourable than in humid zones. In these areas, the water table is subject to pronounced natural fluctuations. Surface recharge methods are best suited to these cases. Coastal dunes and deltaic areas are also often very favourable areas for artificial recharge schemes. Dense urban and industrial concentrations in such areas may render artificial recharge schemes desirable, generally using subsurface recharge wells to inject surface water into the aquifers.

When the quantity and availability of recharge water is highly variable, such as in an intermittent stream, any of the surface application methods are suitable. Basin and pit techniques have the greatest advantages because they can be designed to accommodate expected flood flows. In contrast, shafts and wells have little storage capacity and, therefore, require a more uniform supply of water. Indirect methods, such as induced recharge, are virtually unaffected by changes in surface water flows because the rate of recharge is controlled by extraction rates.

The physical, chemical and biological quality of recharge water also affects the selection of recharge method. If suspended solids are present, surface application techniques tend to be more efficient than subsurface techniques where they can result in clogging of injection wells. It is also important that the recharge water be chemically compatible with the aquifer material though which it flows and the naturally occurring groundwater to avoid chemical reactions that would reduce aquifer porosity and recharge capacity.

Toxic substances must not be present in the recharge water unless they can be removed by pretreatment or chemically decomposed by a suitable land or aquifer treatment system. Similarly, biological agents, such as algae and bacteria, can cause clogging of infiltration surfaces and wells, limiting the subsequent use of the recharged water.

Effectiveness of the Technology

Various artificial recharge experiments have been carried out in India by different organizations, and have established the technical feasibility of the artificial recharge of unconfined, semi-confined and confined aquifer systems. However, the most important, and somewhat elusive, issue in determining the utility of this technology is the economic and institutional aspects of artificial groundwater recharge. Experiences with full-scale artificial recharge operations in India and elsewhere in Asia are limited. As a consequence, cost information from such operations is incomplete.

The available data, from certain hydrological environs in which recharge experiments have been initiated and/or are in progress, suggest that the cost of groundwater recharge can vary substantially. These costs are a function of availability of source water, conveyance facilities, civil constructions, land, and groundwater pumping and monitoring facilities.

Advantages

As surface water augmentation methods, such as dams and diversions, have become more expensive and less promising in terms of environmental considerations, the prospects of storing surplus surface water underground and abstracting it whenever and wherever necessary appears to be more effective technology. In urban areas, artificial recharge can maintain groundwater levels in situations where natural recharge has become severely reduced.

Disadvantages

There are a number of problems associated with the use of artificial recharge techniques. These include disadvantages related to aspects such as recovery efficiency (*e.g.*, not all of the added water may be recoverable), cost effectiveness, contamination risks due to injection of recharge water of poor quality, clogging of aquifers, and a lack of knowledge about the long term implications of the recharge process. Hence, careful consideration should be given to the selection of an appropriate site for artificial recharge in a specific area.

Cultural Acceptability

Cultural considerations, stemming from socio-economic concerns, often enter into the selection of a recharge method and site. The availability of land, land uses in adjacent areas, public attitudes, and legal requirements generally play a role in defining the acceptability of artificial recharge in a given setting.

In urban areas, where land availability, costs and uses in adjacent areas may pose restrictions, injection wells, shafts or small pits requiring highly controlled water supplies and little land area may be preferable to larger scale, surface spreading recharge methods. Surface recharge facilities generally require protected property boundaries, regular maintenance, and continuous surveillance if they are to be acceptable to the public.

Further Development of the Technology

The recharge process is extremely complex, and, due to the numerous factors affecting the process, is only partly understood. The studies on artificial recharge techniques are mostly site-specific and descriptive in nature, which gives little insight into the potential success of implementing this technology in other locations.

Thus, there is a need for further research and development of artificial recharge techniques for a variety of conditions. In addition, the economic, managerial and institutional aspects of artificial recharge projects need to be studied further.

RAINWATER HARVESTING IN PONDS

TECHNICAL DESCRIPTION

The salinity of both surface and ground waters is beyond the tolerable limits for human and agricultural uses, including watering of animals as well as plants. From time immemorial, pond water, replenished by rainfall, is used for drinking and other purposes, including for fish cultivation.

The ponds are specially constructed with bunds so that surface water cannot enter the ponds. This feature keeps the bed of the pond relatively free of sediment. Traditionally, trees have been planted on all four sides of the ponds to provide additional protection of the pond area. In the Sunderban region, animals are permitted to drink from the ponds during the night.

EXTENT OF USE

In southern portion of Hatya Island and other remote islands in the Bay of Bengal, rainwater, collected in ponds, meets nearly 80 per cent of the drinking water requirements during the monsoon season.

The water levels in the ponds fall during the dry season, and the water becomes turbid and more saline due to the deposition of marine aerosols, but the water continues to be used by the local people who have no other source of water, although some households bring water from distant areas.

The pond water is used for bathing, washing of clothes and other household purposes through out the year.

Where tubewells exist, women are generally carry the water from the wells to the point of use. In contrast, even in areas where religious norms are strictly adhered to, men may be found to carry water from the ponds.

Operation and Maintenance

Maintenance of the ponds involves repairing the bunds, cleaning and removing sediment, and general upkeep of the surrounding area.

Level of Involvement

This technology is generally implemented at the community level.

Costs

Capital costs depend on the materials and methods used in constructing and lining the ponds. Typical pond construction costs range between \$350 and \$500 for a 15 m × 15 m pond. Maintenance costs are typically about \$12 per year.

Effectiveness of the Technology

The ponds fill with rainwater during the monsoon season, which, on Hatya Island, extends from April to August. At the conclusion of the monsoons, the water level in the ponds falls rapidly during January and February, and the water becomes salty.

Nevertheless, a large number of people depend on the ponds for their domestic water supply. Hence, this technology appears to be very effective in meeting local needs.

Suitability

This technology is suitable in areas with abundant rainfall to fill the ponds naturally.

Advantages

This technology is relatively inexpensive and can be implemented using local technological skills and community cooperation.

Disadvantages

There is a potential for the pollution of the pond water by animals. Also, if the ponds are poorly maintained, there may be seepage losses. Because the ponds are open, they are subject to evaporative losses.

Cultural Acceptability

Some communities do not drink water from this type of storage pond because it is considered unholy and contaminated.

Further Development of the Technology

This technology may be considered to be fully developed. However, cost effective methods need to be introduced to reduce water losses and pollution to improve its effectiveness.

OPEN SKY RAINWATER HARVESTING

TECHNICAL DESCRIPTION

The open sky rainwater catchment is the least sophisticated form of rainwater harvesting, requiring minimal levels of investment to install, maintain and operate. For this reason, open sky rainwater harvesting is usually practised by relatively poor people who cannot afford appropriately-sized gutters or motkas in which to store water.

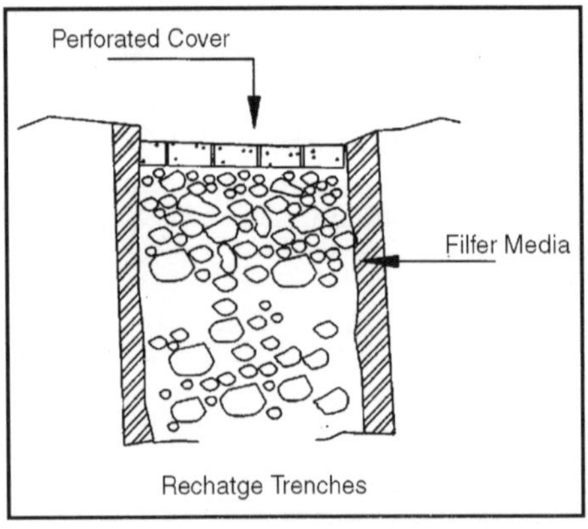

Fig. Open Sky Rainwater Harvesting.

When applied for agricultural use, the polyethylene sheeting may be placed in the open and slanted so that rainwater collecting on the plastic flows off the sheeting and downslope to the agricultural.

Mosquito nets, bed sheets or even sarees (women's garments), inverted umbrellas, and open drums have been used as catchments to harvest rainwater, while fishermen and mariners living offshore for a considerable periods have traditionally harvested rainfall to replenish potable water supplies on their boats.

Operation and Maintenance

This technology requires little maintenance, except that devoted to the periodic cleaning of water storage pots, and few skills to operate.

Level of Involvement

This technology may be implemented at the household level.

Costs

Use of this technology can be had for a minimal investment in materials. In its relatively sophisticated version, that constructed of corrugated iron sheeting, typical costs would be between $5 and $9 for two corrugated iron sheets.

In its least expensive form, that constructed of plastic sheeting, the cost would be about $0.50. Likewise, the cost of storing the harvested rainwater depends on the type of storage vessel used.

Typical costs would range from about $0.25 to over $500: pitchers would cost about $0.25 each; small earthen jars (motkas), $1.20; medium-sized earthen jars, $5.00; large earthen jars, $10.00; cement underground storage tanks, $50 to $500; and, cement above ground storage tanks, $100 to $300.

Effectiveness of the Technology

This technology is effective in meeting the short term water requirements of people living in boats and of small families lacking access to other water supplies.

Suitability

Fig. Woman Taking Fresh Water from a Pond on Hatya Island.

This technology is limited to very small scale uses. Advantages This technology is simple to use, requiring few skills and minimal materials.

Disadvantages

It may be difficult to securely anchor the rainwater catchment materials during windy conditions or during storms, limiting the ability of the technology to harvest rainwater. Because of the limited surface area of the rainwater catchment, this technology does not provide a reliable source of water, and its use is restricted to the wet season.

Cultural Acceptability

There are no known cultural problems associated with the use of this technology.

EXPLOITATION OF WATER SUPPLY

The Vrana Lake is situated on the island of Cres, belonging to the group of Croatian islands of the Adriatic Sea - a marginal sea of the Mediterranean (Southern Europe). With its average area of approximately 5,75 sq.km. and volume of approximately 220 million cu.m. of water of exceptionally high quality on otherwise dry area of the island of Cres (the total area of the island is only 405,78 sq. km.), it represents an invaluable resource in both the environmental and economic sence.

The lake is a kryptodepression, with the water level between 16,70 m a.s.l. and 9,11 m a.s.l. on the average 13,13 m a.s.l. in the period from 1929 to 1995. The lake bottom is at 61.3 m below the mean sea level. The average annual rainfall in the area of the Vrana lake is 1064 mm, the mean annual air temperature is +14.8°C, and the annual evaporation from free water surface is 1161 mm. The falling trend of the water level in the lake in the eighties, never recorded before, caused a great concern among proffesuinals and the public regarding the future of the lake, and initiated the beginning of complex research work. In the recent years this trend has been stopped; however the need for protection of the lake is still present in the sense of defining the regime of water pumping from the lake. Because of its position and size, the issue of the origin of the lake water has been attracting the attention of scientists for a lond time. Already in the past century contrary assumptions appeared regarding its recharging - from the mainland or the local island catchment area. The results of all hydrological investigations carried out so far, both earlier and recent, althought not identical in figures and conclusions, speak in favour of the latter assumption. Mechanism of lake's aquifer functioning in way of hydraulic relation between salt and fresh water was analysed in Ozanic and Rubinic. In Bonacci hypotetical hydraulic model with assumption Lake's recharging from the mainland was analysed. The results of these hydraulical investigations confirm hydrological conclusions, so the problem of functioning and protection of Vrana Lake must be consider as part of island karst aquifer *i.e.* balance relation between salt and fresh water in according to the Ghyben-Hertzberg Low.

PROPERTIES OF THE VRANA LAKE SYSTEM

However, due to the specific properties of the Vrana Lake which has no directly measurable imflow nor run-off, the hydrological analyses of functioning of the lake system based on measuring of water balance parameters of the lake system (water level monitoring, precipitation regime, water pumping from the lake, evaporation from lake surface) have made a considerable contribution to the general level of knowledge on this natural phenomenon and its protection.

Due to its size compared to the comparatively small island enviroment, the lake containing 220 million cu.m. of potable water of exceptionally high quality affects significantly the dynamics of functioning of the entire aquifer. Except its position and size, the Vrana Lake is specific due to the fact that inflow into the lake, as well as the outflow, is accomplished by so far unidentified underground ways. No significant permanent springs have been noticed on the lake, except two temporary springs with the yield up to 0.005 cu.m./sec. Compared to the mater in the mainlend area beyond the direct influence of the sea, salinity of the lake water is somewat higher, ranging between 62 and 92 mgl-1. In the wider zone of the Vrana Lake - on the coast of the island of Cres several springs, vrulja and coastal sources have been noticed, as well as a zone of diffused fresh water outflow. Due to the constant outflow regime, some of these phenomena may be related to the Vrana Lake, or its aquifer, as possible privilleged directions of its discharging. Water level observations have been carried out on a daily basis since 1928, completed by rainfall observations on the catchment area and directly along the shore.

An expressed trend of water level decrease in the lake is noticed, in particular in the eighties. The trend of water level decrease in the entire period of analysis, 1929 to 1995, was 0,04 m per annum, while in the period from 1985 to 1990 it was even 0,48 m per annum.

This trend caused the concern of the public and of the professionals for the destiny of the lake and possible disturbance of its balance, and possible breakthrough of sea water into its system. On the other hand, it may be seen that water pumping from the lake, which started in 1952, shows a clear growing trend, with the annual average for the 1964 to 1995 period of about 65 000 cu.m. per annum. During the recent decade, the average pumping rate from the lake is 0,072 cu.m./sec., with the maximum during the summer season, up to 0,160 cu.m./sec. Fortunately, such drastic trend of water level decrease that could have endangered the balance of the fresh water aquifer of the Vrana Lake, was not continued - the rainfall conditions wereb improved, and due to the war situation in Croatia in the early nineties, the touristic traffic was reduced, resulting in reduced water pumpings from the lake, compared to pre-war maximum values.

RESULTS OF HYDROLOGICAL ANALYSIS

Multiple regression analysis of the annual changes of the lake status, in dependance on the magnitude of relevant hydrological parameters for the period from 1980 to 1995, when the most complete input data were available, gave a very acceptable regression dependance with the linear correlation coefficient k=0,96, wich may be seen. The resulting equation is: Hn=2,7369 On+1,12836 In–4,0069 Cn+0,23208 Gn-0,31284 H(n–1)+0,3718 (1) Hn - annual water level change, On - annual rainfall, In-annual evaporation, Cn-annual pumping, Gn-calculated losses through sinking determinated on the basis of the sinking curve and H(n-1)-mean annual water levels in the previous year–all expressed in (m) with regard to the mean lake level.

Such analyses, along with detailed analyses of sinking from the lake system and, provided the preconditions for elaboration of the complex mathematical model of functioning of the lake system of RANA which was used to compute the mean monthly inflows into the Vrana Lake. According to the computations, in the period from 1929 to 1995, the mean annual inflow into the Vrana Lake system was 0,588 cu.m./sec., and its value varied from 1,144 cu.m./sec. and 0,273 cu.m./sec.

The average value of inflow from the direct catchment area was calculated as 0,393 cu.m./sec., and figure of 0,195 cu.m./sec. refers to the rain falling directly on lake surface. The area of the orographic catchment area sourrounding the Vrana Lake oscillations and to the analysis of regional run-off, it comes out that the average catchment area satisfying the Vrana Lake recharging balance is only about 24 sq.km. This size of catchment area is not clearly determinable, changing due to the karst character of the terrain in dependance on hydrological conditions.

LAKE PROTECTION FROM OVERPUMPING

As already mentioned, the Vrana Lake is the only source of water supply of the islands of Cres and Loiknj, and at the present requirements the annual quantity of water pumped out of the lake reaches 2,3 million cu.m.

As the calculated annual average inflows are 18.5 million, and the evaporation losses at the mean water level are of the order of 6,7 million cu.m., and the losses through sinking from the lake system are 11,7 cu.m. it is obvious that pumping, although globally the smallest element of the water balance, disturbs the balance of the Vrana Lake system and influence lowering of the water level.

This lowering of the mean lake level occurs to the case when the sinking losses are reduced in the magnitude of the increase of the pumped quantity.

Namely, a functional relation has been determined between the quantity of sinking and the water level in the lake, and its gradient in domain of the observed water levels per each meter is approx. 0,028 cu.m./sec.

Until 1995, the total quantity of water pumped out of the lake reached 42 million m^3 which would, at the mean water level account for 7.3 m. The analysis showed that such water consumption, in spite of the expressed trend of water level lowering, caused, instead of the said 7.3 m, the actual lowering of the lake level by approximately 2 m.

Therefore, it is obvious that this water loss was compensated by the change of hydrological conditions in the lake-first of all by reduction of the losses from the aquifer due to his lower level in relation to the sea, and partly due to the increased recharging from the underground part of the aquifer, caused by lowering of the lake level. The trend of recorded annual water levels in the Vrana Lake, and of values, obtained by mathematical simulation on the model of RANA, assuming there was no pumping, but also under the assumption that in the period of analysis pumping was even more intensive than the actual mean annual rate of 0,072 cu.m./sec., *i.e.* that they were double - 0,150 cu.m./sec., and 0,250 cu.m./sec., which is equal to the earlier forecasts of possible maximum allowed pumpings rates.

Althought the simulations of the Vrana Lakeb behaviour in cases of its intensified use have carried out on the basis of extrapolation of the determined regularities of behaviour in the real - higher water levels in the lake, and are therefore largely approximative, they still show that more intensive pumping of water from the lake may result in excessive consumption which, in turn, could cause disturbance of its dynamic balance with the sea. According to analysis minor planned increases of the pumping rate of the order of 20 to 30 per cent for the needs of the island population are not yet alarming.

The formed fresh water lense of the island aquifer, by its higher pressure, on the principles of the Ghyben-Hertzberger Low, maintains balance with the sea and prevents its breakthrough into the lake. As the bottom of the Vrana Lake is even 61 m below the mean sea level, and with respect to the theoretical balance ratio (1m of fresh water lense above the sea level corresponds to 40 m of fresh water below the sea level) in that case the breakthrough of sea into the lake would occur at the lake level of 1,5 m a.s.l.

Such breakthrough into the lake would be a disaster for the lake system because the thermal currents in the lake, as well as winds, would cause salination of the entire lake, without any possibility of desalination later on. However, in reality, as the Vrana Lake is in the karst environment the limit of the contact of fresh water and the sea is not distinct, but involves a wider fresh and sea water mixing zone, which is considerably closer to the lake.

Therefore, the above-mentioned theoretically possible lowering of the lake level should not be allowed, and increased pumping of water from the lake must be subject to constant control, and limited to the intensity that will not endanger the balance of the lake.

Vrana Lake is part of the complex system of island's karst aquifer, which functions in accordance with the principals of dynamical balance of the fresh water and the sea.

In this chapter are given the main characteristic of its hydrological system as also the analysis of problems of pumping from the lake.

This chapter shows that we can analysed aquifer behaviour by hydrological modeling when their natural hydraulic give limitation of excessively exploitation. It has been determined that the observed lowering of the water level in the Lake during the eighties was the result of the coincidence of extremely dry hydrological conditions and increased consumption of water from the Lake (pumping).

Existing pumping of average 0,072 m^3/s does not disturb Vrana Lake to its exhaustion. Such development, in perspective, with unavoidable increase of water consumption standards, results in new requirements for water pumping out of the lake, thereby bringing new risks of disturbing the existing water quality.

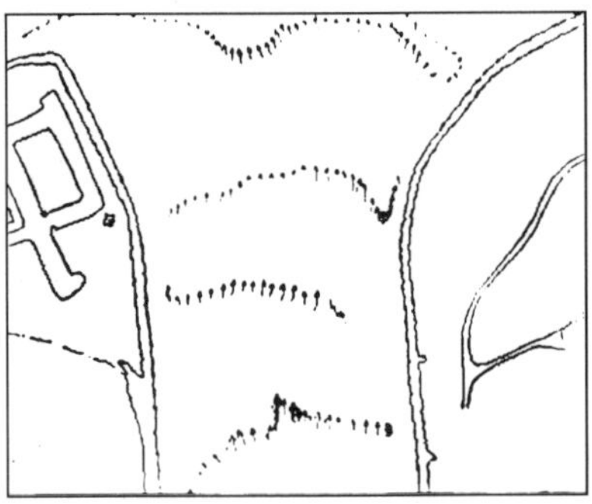

Fig. Location Map.

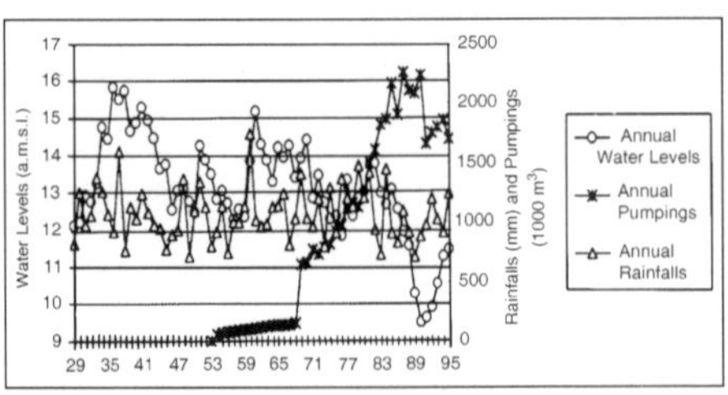

Fig. Annual Values of Rainfall, Water Water Levels andWater Pumpings from the Vrana Lake.Levels

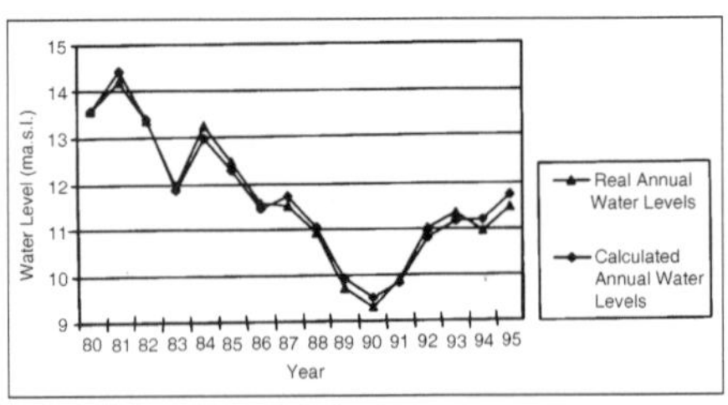

Fig. Ratio of Recorded and Calculated Water Levels According to Multiple Regression Analysis

In accordance with the conceptions we have found out it is obvious that the former estimations of permitted 0,250 m3/s were too optimistic. It has been established that rapid trend of falling of water level was caused by more intensive pumping from Vrana Lake and driver hydrological conditions.

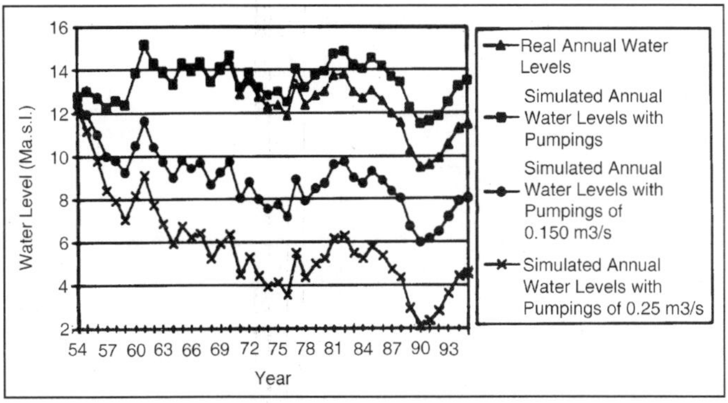

Fig. Annual Water Levels in Real Conditions, Withought Pumping, and With Pumping of 0,150 m^3s^{-1}, and 0,250 m^3s^{-1}.

HYDRAULIC PROCESSES ANALYSIS SYSTEM

HyPAS builds on the inherent power of GIS while supplying easier tools for facilitation of hydraulic process analysis and reducing learning time typical with GIS implementation. HyPAS is an extension to ArcView, a commercially available software package marketed by Environmental Systems Research Institute (ESRI), and requires an additional extension from ESRI, 3D Analyst. HyPAS was designed for the non-GIS expert with ease of use as a priority. The HyPAS was designed to perform all major functions after collection of data to report writing.

HyPAS's velocity analysis tools cover three basic applications: contouring an area in plan view from a user-defined constituent and depth range, generating cross sections from a transect, and plotting vector magnitude and direction in plan view from a user-specified depth range. HyPAS's sediment sample analysis tools allow the user to generate frequency weight plots, calculate composite sample plots, and perform varied analysis routines. HyPAS's project management tools allow the user to import photographs for project enhancement and import time series data to manage and plot.

GENERATING PLAN VIEW CONTOURS

HyPAS provides the capability of generating contours in plan view for the different constituents collected from the Acoustic Doppler Current Profiler (ADCP) instrument. HyPAS will create colour-shaded contours and contour lines. This tool allows the user to calculate the depth average of the specific values from the entire water column or a specific depth range. The user selects the survey data to contour by drawing a box or polygon around them. A depth range is specified, and the user must choose which constituent to contour.

HyPAS then calculates the contour from the depth range specified. An example plan view contour of total velocity magnitude was created from ADCP data in the Columbia River, Oregon. The yellow dots represent actual velocity profiles collected by the ADCP as the boat moves along the line. The legend for the colour-shaded contours represents velocity magnitude in ft/sec. The units for velocity magnitude depend on how the data were processed. Often, velocity magnitude is represented in cm/sec.

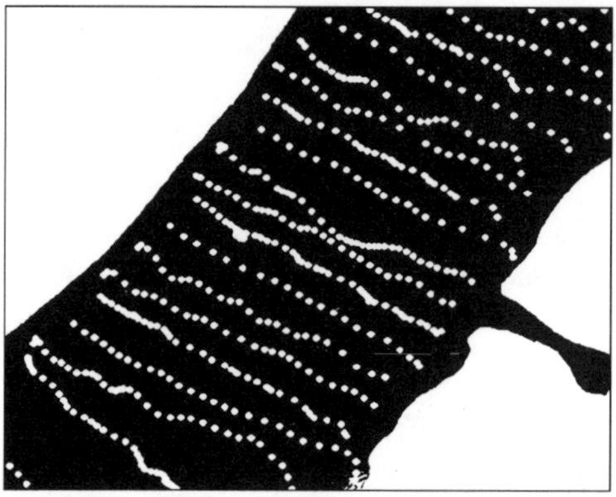

Fig. Plan View Contour of the Total Velocity Magnitude from an ADCP Survey in the Columbia River.

GENERATING PLAN VIEW VECTOR

HyPAS provides the capability of generating a vector plot of magnitude and direction in plan view for the velocity data collected from the Acoustic Doppler Current Profiler (ADCP) instrument. HyPAS will create a plot with vectors pointing in the direction of flow and scaled in size by magnitude. This tool allows the user to calculate the depth average of velocity magnitude and direction from the entire water column or from a specific depth range. The user selects the survey data points to plot by drawing a polygon around them. A depth range is specified. HyPAS then calculates and plots the depth-averaged magnitude and direction for those survey points. Figure below is an example of depth averaged velocity magnitude and direction vectors from ADCP data at Shinnecock Inlet, New York.

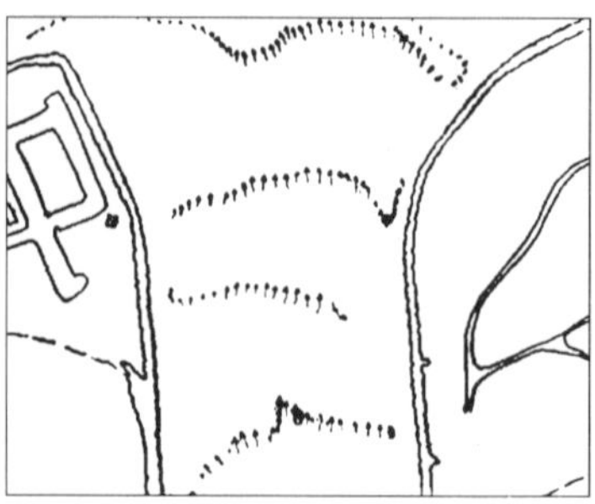

Fig. Plan View Velocity Magnitude and Direction Plot from Shinnecock Inlet, New York (Flood Current).

GENERATING CROSS SECTIONS

HyPAS provides the capability of generating a cross section from a specific transect or points along multiple transects from velocity survey data. This tool allows the user to calculate a cross section from any constituent in the data. This includes the north, east, and vertical components of velocity as well as acoustic backscatter. The user selects the survey data points by drawing a polygon around the individual points and selecting a starting point.

The constituent to contour is selected. HyPAS then interpolates and plots the cross section. A cross section was created from ADCP data collected from an area called Victoria Bend in the Mississippi River.

IMPORTING AND DISPLAYING IMAGES

HyPAS provides the capability of importing and displaying digital photographs and images to enhance project management. This tool is not designed for image processing, but rather as an end product aid for representing analysis results and methods. This tool allows the user to import digital photographs and images and attach them to specific spatial locations throughout the study area. The images are stored using a separate image theme as shown in Figure. On the image theme, a small solid filled circle with an "i" denotes the locations of the images. Presently, a maximum of ten digital images can be stored at each spatial location. This tool is useful for recording conditions at a site or progress of a construction site.

Images allow the user the opportunity to store and display instruments, flow conditions, historic photos, and other visual information quickly and easily for reference. Several photographs were added to a GIS project for Shinnecock Inlet, New York. Two photographs are displayed showing the inlet at previous times in its history.

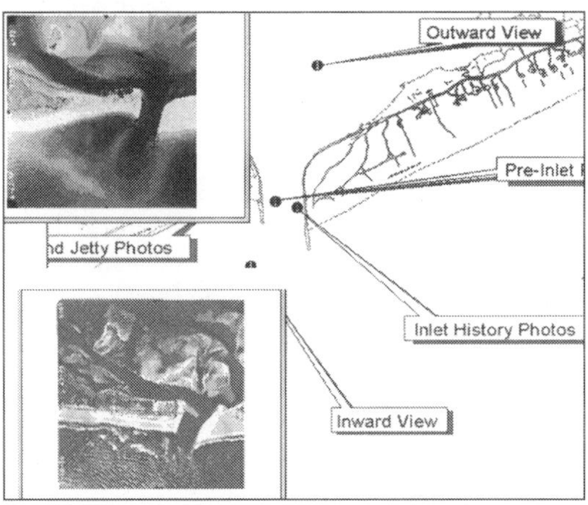

Fig. Shinnecock Inlet GIS Project Utilizing HyPAS Tools for Displaying Photographs.

TIME SERIES ANALYSIS

HyPAS provides capabilities for importing, storing, analyzing, and exporting time-series data. This is a project management tool that allows the user to have time-series data flags throughout his project. The user create a location for time–series data and import the data for that location.

Later the user can choose that location and HyPAS will display all the data types, which have been imported. HyPAS will plot the data on an X, Y plane using the plot and axis information given by the user. HyPAS automatically exaggerates the Y axis to fit an 8.5 x 11 landscape plot. Figure shows a time-series plot of wind speed information.

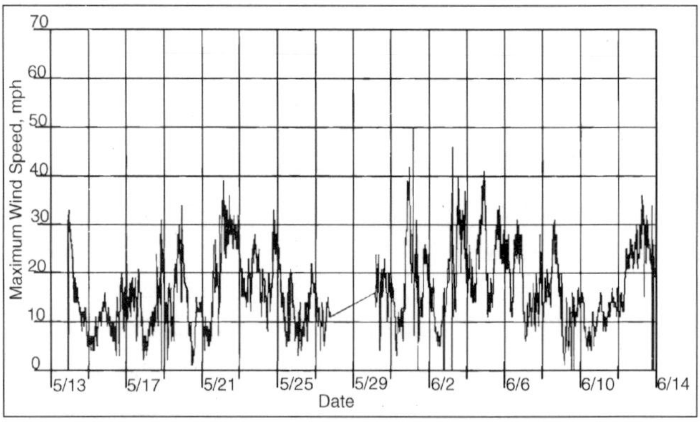

Fig. Maximum Wind Speed Time-Series Data Plot.

PLOTTING GRAIN SIZE DISTRIBUTION

HyPAS provides the capability of plotting grain-size distribution for sediment sample data. The user can plot cumulative frequency weight or frequency weight per cent histograms. HyPAS can plot all samples selected or calculate a composite for the selected samples. The user has the option of sorting the tabular data before making a selection of samples for display in the plotting window.

Once sediment samples are imported, the user must select the samples desired for a grain-size distribution plot before selecting the Sediment Sample Tool. The option to plot the cumulative frequency weight percents, the frequency weight per cent histogram, or a composite sample plot is provided.

HyPAS then generates the distribution plot. Figure shows frequency weight per cent histograms with a composite histogram from a subset of sediment samples collected from the Shinnecock Inlet.

A cumulative frequency weight per cent plot of the same sample is also shown. The thicker line delineates the composite sample. Various statistical parameters describing this composite sample are stored to a table. These statistics include the median, other percentiles, the Fill Factor (RA), and the Renourishment Factors (RJ). Different composites can then be easily compared.

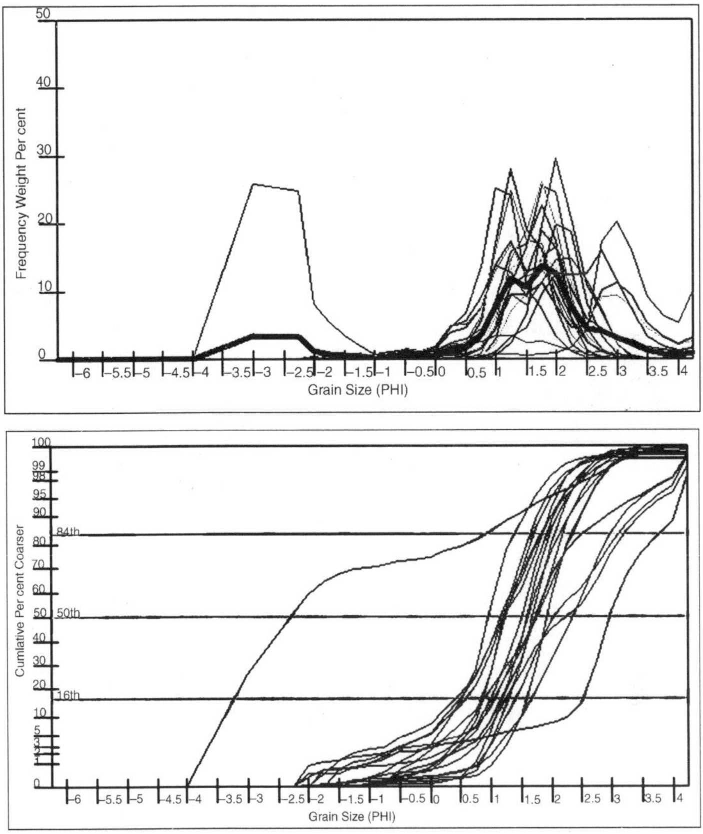

Fig. Example of a Series of Sediment Samples and Composite Sample. The First Plotdisplays the Frequency Weight Per cent Histograms, and the Second Plot Displays the Cumulative Weight Per cent.

AGRICULTURAL WATER USE

The images of seemingly endless crop fields of the American Midwest and the lush San Joaquin Valley of central California are powerful symbols of the agricultural might of the United States. In the past century, the United States has become the greatest producer of food in the world. Water has always been a vital part of agriculture. Just like humans, crops need water to survive and grow. The process where dry land or crops are supplied with water is called irrigation. A century ago, the relatively small fields of a local farmer in many areas of the United States could receive enough moisture from rainwater, along with water that could be diverted from local streams, rivers, and lakes.

The growth of huge corporate farms that are thousands of acres in size has taken the need for water to another scale. For these operations, water needs to be trucked in, pumped up from underground, and obtained from surface water (freshwater located on the surface) sources in large quantities. In modern times, in countries such as the United States and Canada, agriculture is not the largest user of water but is the largest consumer of water. Other activities such as the oil industry use more water than does agriculture. But, in these other industries, much of the water is put back into the ground or surface water after being used. Agriculture consumes water; the water does not go back to the surface or to the groundwater.

USES OF WATER IN AGRICULTURE

There are four main areas of water use in agriculture: growing of crops, supplying drinking water to livestock, cleaning farm buildings and animals, and supplying drinking water for those who work on the farm. The amount of each category varies just as to the type of farm. For example, farms in the eastern part of North America usually receive enough rainfall and water from melting snow to meet most of the water needs. But drier areas, such as the U.S. and Canadian prairies, regions of Mexico, and some mountainous regions of the West do not receive sufficient natural moisture. On these farms, water must be supplied through irrigation.

IRRIGATION

Nearly 60 per cent of the world's freshwater that is used by humans is used for irrigation. Of this water that is applied to crop fields, only about half returns to surface water or groundwater sources. The rest is lost by natural processes such as evaporation (when liquid water changes to water vapour) and transpiration (when water from plant leaves is transformed into water vapour), and accidental occurrences such as leakage from pipes or spillage. There are various methods to irrigate crops. The oldest, "low-tech" way is to flood the field. This flood type of irrigation has been used for centuries and remains popular for crops like rice. Field flooding is very wasteful, since only about half of the water used actually gets to the plant.

The efficiency of flood irrigation can be improved by making the land contoured, such as eliminating small hills and putting steps (terraces) on larger hills to prevent water from flowing over certain portions of the field and gathering in another part of the field. The flooding of a field can also be controlled by releasing water from dams (barriers) alongside the field, adding water to the field only when needed. Water that flows off of a field can be captured in a pond and re-used. A newer and much more efficient technique of water use is called drip irrigation.

In drip irrigation, water runs through pipes that have tiny holes in them. When buried underground, water can ooze out of the pipe into the soil near the roots of the plants. The loss of water is reduced and less water is required to grow the crops. A popular means of irrigation is spraying. Water flows through a tube and is shot out through a system of spray nozzles positioned along the length of the tube. The tube can be fixed in one position or can be moved manually or automatically from place to place.

A visual example of a spray irrigation system is a green circle seen from an airplane passing over farmland. The green circles are crops that are being irrigated by a circular sprayer. Spray irrigation is sometimes wasteful, as water that is sprayed can evaporate or be blown away before hitting the crop. Some farmers now use an irrigation method where water is gently sprayed from pipes that are suspended over the crop. This method allows about 90 per cent of the water to reach the crop. With the knowledge that surface and groundwaters are resources that can be overused, agricultural scientists and modern farmers are paying attention to methods of conserving and re-using water while maintaining the growth of their crops.

6

Application of Remote Sensing in Precision Agriculture

SOIL SURVEYS REMOTE SENSING

Soil maps are also sometimes used to determine management zones. Soil maps are becoming part of the GIS database. The grid sampling technique takes separate soil samples from uniform sized grids laid out over the field. A problem with this type of sampling is the variability that can exist in soil types with in each grid. This variability makes it much tougher to determine soil characteristics within the grid for crop input management purposes. To minimize this problem smaller grids are required which then requires many more soil samples to be take for a larger number of grids. Soil samples can become a major cost of precision farming. An alternative to grid sampling is targeted or zone sampling. The soil samples are located in homogeneous management zones instead of uniformly spaced grids. The zones are laid out using a process similar to computer based unsupervised image classification. Images obtained from multispectral remote sensors are taken of the vegetated areas of the field. The pixel digital numbers for each band are separated into statistically separable clusters that are classified into homogeneous zones. This cuts down on the soil, terrain, plant growth, and other variability within each area to be managed; thus fewer soil samples are needed for each area.

Except for county soil surveys remote sensing has not gained wide acceptance as a mapping tool for soil characteristics. This is because "the reflectance characteristics of the desired soil properties (*e.g.*, organic matter, texture, iron content) are often confused by variability in soil moisture content, surface roughness, climate factors, solar zenith angle, and view angle".

DRAINAGE MAPS

Subsurface drainage tile lines that have been installed, as long as 50 or more years ago are still partially or totally functional today. Often the existence or location of older tile lines has been lost as landowners die or sell their property. Some states, including Iowa, are now starting to require landowners to prepare and file drainage plats with county recorders when new tile lines are installed.

It is desirable to have accurate drain tile maps for maintenance purposes or for the installation of new additional tile line systems. Installation of new tile lines may cut through old tile lines at unknown locations. Building livestock manure lagoons, which cut through old unknown and uncharted tile lines may cause environmental damage from manure leaking through the old tile lines. Colour infrared ("CIR") aerial photographs have been shown to be an effective tool in locating unknown subsurface tile lines. The image data is digitized for preprocessing and then geo-referenced using ground control points. The CIR photographs show different tones of gray depending on soil type and moisture. By filtering out spectral reflectance differences due to soil type, soil moisture content in dry soils that have a higher reflectance can be identified from lower reflectance wet soils. The resulting image shows were the tile lines are located and whether they are working properly.

Normal colour aerial photographs can also be used to locate tile lines. Simple colour photographs offer tile line images similar to CIR but at a lower cost. If the soil is too dry such, the tile lines will not be visible in the image. The images similar, must be acquired when the soil is bare and within a few days after an adequate rain. High resolution and on demand temporal availability make images acquired from aircraft platforms ideal for acquiring this kind of image data.

Variable Rate Technology

One method of controlling variability within the field is VRT. VRT allows the grower to apply the quantity of crop inputs needed at a precise location in the field based on the individual characteristics of that location. Crop inputs that can be varied in their application commonly include tillage, fertilizer, weed control, insect control, plant variety, plant population, and irrigation. Typical VRT system components include a computer controller, GPS receiver, and GIS map database. The computer controller adjusts the equipment application rate of the crop input applied.

The computer controller is integrated with the GIS database, which contains the flow rate instructions for the application equipment. A GPS receiver is linked to the computer. The computer controller uses the location coordinates from the GPS unit to find the equipment location on the map provided by the GIS unit. The computer controller reads the instructions from the GIS system and varies the rate of the crop input being applied as the equipment crosses the field. The computer controller will record the actual rates applied at each location in the field and store the information in the GIS system, thus maintaining precise field maps of materials applied.

Although VRT can control inputs applied to crops, it cannot control factors such as soil type, weather climate, and topography that are fixed.

Monitor Crop Health

Remote sensing data and images provide farmers with the ability to monitor the health and condition of crops. Multispectral remote sensing can detect reflected light that is not visible to the naked eye.

The chlorophyll in the plant leaf reflects green light while absorbing most of the blue and red lightwaves emitted from the sun. Stressed plants reflect various wavelengths of light that are different from healthy plants. Healthy plants reflect more infrared energy from the spongy mesophyll plant leaf tissue than stressed plants. By being able to detect areas of plant stress before its becomes visible, farmers will have additional time to analyze the problem area and apply a treatment.

Water Stress

The use of remote sensors to directly measure soil moisture has had very limited success. Synthetic Aperture Radar ("SAR") sensors are sensitive to soil moisture and they have been used to directly measure soil moisture. SAR data requires extensive use of processing to remove surface induced noise such as soil surface roughness, vegetation, and topography.

A crop evapotranspiration rate decrease is an indicator of crop water stress or other crop problems such as plant disease or insect infestation. Remote sensing images have been combined with a crop water stress index ("CWSI") model to measure field variations.

Simple panchromatic aerial photographs have been used to spot irrigation equipment problems. Strips in the vegetation images point to problems with water application rates from defective water nozzles.

Weed Management

One goal of precision farming is to cut crop production inputs, which result in cost and environmental savings. Conventional farming methods apply herbicides to the entire field. Site-specific variable-rate application puts the herbicide where the weeds are. Aerial remote sensing has not yet proved to be very useful in monitoring and locating dispersed weed populations. Some difficulties encountered are that weeds often will be dispersed throughout a crop that is spectrally similar, and very large-scale high resolution images will be needed for detection and identification.

The use of machine vision technology systems to detect and identify weeds places remote sensors directly on the sprayer equipment. Being close to the crop allows for very high spatial resolutions.

Machine vision systems have the ability to be used in the field with the real-time capabilities that are necessary to control sprayer equipment.

Insect Detection

Aerial or satellite remote sensing has not been successfully used to identify and locate insects directly. Indirect detection of insects through the detection of plant stress has generally not been used in annual crops. The economic injury level for treatment is usually exceeded by the time plant stress is detected by remote sensing. Entomologists prefer to do direct in field scouting in order to detect insects in time for chemical treatments to be effective and economical.

Nutrient Stress

Plant nitrogen stress areas can be located in the field using high-resolution colour infrared aerial images. The reflectance of near infrared, visible red and visible green wavelengths have a high correlation to the amount of applied nitrogen in the field. Canopy reflectance of red provides a good estimate of actual crop yields.

Yield Forecasting

Plant tissue absorbs much of the red light band and is very reflective of energy in near infrared ("NIR") wavebands. The ratio of these two bands is referred to as the vegetation index ("VI"). The difference of red and NIR measurements divided by their sum is normalized difference VI ("NDVI"). For crops such as grain sorghum, production yields, leaf area index ("LAI"), crop height and biomass have been correlated with NDVI data obtained from multispectral images. In order to get reasonably accurate yield predictions this data must be combined with input from weather models during the growing season.

PRECISION FARMING SYSTEM

Precision farming basically depends on measurement and understanding of variability, the main components of precision farming system must address the variability. Precision farming technology enabled, information based and decision focused, the components include, (the enabling technologies) Remote Sensing (RS), Geographical Information System (GIS), Global Positioning System (GPS), Soil Testing, Yield Monitors and Variable Rate Technology. Precision farming requires the requisition, management, analysis and output of large amount of spatial and temporal data. Mobile computing systems were needed to function on the go in farming operations because desktop systems in the farm office were not sufficient. Because precision farming is concerned with spatial and temporal variability and it is information based and decision focused. It is the spatial analysis capabilities of GIS that enable precision agriculture. GPS, DGPS has greatly enabled precision farming and of great importance to precision farming, particularly for guidance and digital evaluation modelling position accuracies at the centimeter level are possible in DGPS receivers. Accurate guidance and navigation systems will allow for farming operations at height and under unfavorable weather conditions even.

In India, we have all these technologies available and they can be implemented through agricultural training centers by giving training to agriculture officers in these technologies.

BASIC STEPS IN PRECISION FARMING

The basic steps in precision farming are,

- Assessing variation
- Managing variation and
- Evaluation.

The available technologies enable us in understanding the variability and by giving site specific agronomic recommendations we can manage the variability that make precision agriculture viable. And finally evaluation must be an integral part of any precision farming system. The detailed steps involved in each process are clearly depicted in a diagramme.

ASSESSING VARIABILITY

Assessing variability is the critical first step in precision farming. Since it is clear that one cannot manage what one does not know. Factors and the processes that regulate or control the crop performance in terms of yield vary in space and time. Quantifying the variability of these factors and processes and determining when and where different combinations are responsible for the spatial and temporal variation in crop yield is the challenge for precision agriculture. Techniques for assessing spatial variability are readily available and have been applied extensively in precision agriculture. The major part of precision agriculture lies in assessing to spatial variability. Techniques for assessing temporal variability also exist but the simultaneous reporting a spatial and temporal variation is rare. We need both the spatial and temporal statistics. We can observe the variability in yield of a crop in space but we cannot predict the reasons for the variability. It needs the observations at crop growth and development over the growing season, which is nothing but the temporal variation. Hence, we need both the space and time statistics to apply the precision farming techniques. But this is not common to all the variability/factor that dictate crop yield. Some variables are more produced in space rather with time, making them more conducive to current forms of precision management.

MANAGING VARIABILITY

Once variation is adequately assessed, farmers must match agronomic inputs to known conditions employing management recommendations. Those are site specific and use accurate applications control equipment. We can use the technology most effectively. In site-specific variability management. We can use GPS instrument, so that the site specificity is pronounced and management will be easy and economical. While taking the soil/plant samples, we have to note the sample site coordinates and further we can use the same for management. This results in effective use of inputs and avoids any wastage and this is what we are looking for.

The potential for improved precision in soil fertility management combined with increased precision in application control make precise soil fertility management as attractive, but largely unproven alternative to uniform field management. For successful implementation, the concept of precision soil fertility management requires that within-field variability exists and is accurately identified and reliably interpreted, that variability influences crop yield, crop quality and for the environment. There fore inputs can be applied accurately. The higher the spatial dependence of a manageable soil property, the higher the potential for precision management and the greater its potential value.

The degree of difficulty, however, increases as the temporal component of spatial variability increases. Applying this hypothesis to soil fertility would support that Phosphorus and Potassium fertility are very conducive to precision management because temporal variability is low. For N, the temporal component of variability can be larger than its spatial component, making precision N management much more difficult in some cases.

EVALUATION

There are three important issues regarding precision agriculture evaluation.

- Economics
- Environment and
- Technology transfer.

The most important fact regarding the analysis of profitability of precision agriculture is that the value comes from the application of the data and not from the use of the technology. Potential improvements in environmental quality are often cited as a reason for using precision agriculture. Reduced agrochemical use, higher nutrient use efficiencies, increased efficiency of managed inputs and increased production of soils from degradation are frequently cited as potential benefits to the environment. Enabling technologies can make precision agriculture feasible, agronomic principles and decision rules can make it applicable and enhanced production efficiency or other forms of value can make it profitable.

The term technology transfer could imply that precision agriculture occurs when individuals or firms simply acquire and use the enabling technologies. While precision agriculture does involve the application of enabling technologies and agronomic principles to manage spatial and temporal variability, the key term is manage. Much of the attention in what is called technology transfer has focused on how to communicate with the farmer. These issues associated with the managerial capability of the operator, the spatial distribution of infrastructure and the compatibility of technology to individual farms will change radically as precision agriculture continues to develop.

Technology Transition

Precision agriculture is dependent on the existence of variability in either or both product quantity and quality. If this variability does not exist then a uniform management system is both the cheapest and most effective management strategy and precision farming is redundant. Thus, in precision farming, "Variability of production and quality equals opportunity". Having said this, the nature of the variation is also important in determining the potential for PA in a system. For example the magnitude of the variability may be too small to be economically feasible to manage. Alternatively the variability may be highly randomized across the production system making it impossible to manage with current technology. Finally the variability may due to a constraint that is not manageable. Thus the implementation of precision farming is limited by the ability of current variable rate technology (VRT-machinery/technology that allows for differential management of a production system) to cope with the highly variable sites and the economic inability to produce returns from sites with low variability using precision farming (VRT).

DEFINITION OF PRECISION AGRICULTURE

"Precision Farming is the title given to a method of crop management by which areas of land/crop within a field may be managed with different levels of input depending upon the yield potential of the crop in that particular area of land. The benefits of so doing are two fold:

- The cost of producing the crop in that area can be reduced and,
- The risk of environmental pollution from agrochemicals applied at levels greater than those required by the crop can be reduced".

Precision farming is an integrated agricultural management system incorporating several technologies. The technological tools often include the *global positioning system, geographical information system, yield monitor, variable rate technology, and remote sensing*.

The *global positioning system* ("GPS") is a network of satellites developed for and managed by the U.S. Defence Department. The GPS constellation of 24 satellites orbiting the earth, transmit precise satellite time and location information to ground receivers. The ground receiving units are able to receive this location information from several satellites at a time for use in calculating a triangulation fix thus determining the exact location of the receiver. A *geographical information system* ("GIS") consists of a computer software data base system used to input, store, retrieve, analyze, and display, in map like form, spatially referenced geographical information.

GOAL OF PRECISION FARMING

The goal of precision farming is to gather and analyze information about the variability of soil and crop conditions in order to maximize the efficiency of crop inputs within small areas of the farm field. To meet this efficiency goal the variability within the field must be controllable.

Efficiency in the use of crop inputs means that fewer crop inputs such as fertilizer and chemicals will be used and placed where needed.

The benefits from this efficiency will be both economical and environmental. Environmental costs are difficult to quantify in monetary terms. The reduction of soil and groundwater pollution from farming activities has a desirable benefit to the farmer and to society.

Research

Precision farming is an integration of several technologies. U.S. and foreign governments originally paid for the development and support of technologies such as GPS, Remote Sensing, and GIS, for military or other civilian purposes long before the emergence of precision farming.

It is the use of these advanced technologies that has generated enormous amounts of data to process with computers. A very basic question that still needs to be answered by researchers is what does all of this data mean in order for farmers to make profitable management decisions? In other words we have the technology to gather the data from the field but we don't yet have the knowledge to transform the data into answers for agricultural management decisions.

Universities around the world, and foreign and U.S. government entities such as the Agricultural Research Stations of the USDA are conducting extensive research on precision farming. Research projects will apply technologies such as remote sensing, GPS, GIS, and VRT to create management decision support systems. A goal of many publicly funded research institutions is to promote technology transfer from government agencies to the private sector.

PRECISION FARMING: DREAMS AND REALITIES FOR INDIAN AGRICULTURE

Ever since the man appeared on the earth, he has been harnessing the natural resources to meet his basic requirements. Reference to soil, water and air as basic resources, their management and means to keep them pure are mentioned in the Vedas, Upanishads and in ancient Hindu literature. The phenomenal increase in population of both man and animal in the last century and fast growing industrialization and urbanization in last few decades have overstrained the natural resource base, which are getting degraded much faster than ever before.

Thus, the attention of whole world is focused on how to increase production to feed the burgeoning population and the question uppermost in every ones mind is "Can we produce enough food in a sustainable manner without damage to the natural resource base?"

Belying all predictions made to the contrary, India could achieve unprecedented increase in the food grain production as a result of expansion of irrigation and technological advancement in agriculture. While it has been a satisfying experience, Indian agriculture would need a new vision to make rapid progress in the ensuing millennium. To achieve the required growth will not be easy as some of the existing production systems are based on unsustainable use of the resources. The signs of fatigue in the natural resources have already appeared which is a cause for serious concern to the planners, decision-makers and researchers alike.

Precision Agriculture is the application of technologies and principles to manage spatial and temporal variability associated with all aspects of agricultural production for improving production and environmental quality.

The success in precision agriculture depends on the accurate assessment of the variability, its management and evaluation in space-time continuum in crop production.

The agronomic feasibility of precision agriculture has been intuitive, depending largely on the application of traditional arrangement recommendations at finer scales.

The agronomic success of precision agriculture has been quite convincing in crops like sugar beet, sugarcane, tea and coffee. The potential for economic, environmental and social benefits of precision agriculture is largely unrealized because the space-time continuum of crop production has not been adequately addressed.

Successful implementation of precision agriculture depends on numerous factors, including the extent to which conditions within a field are known and manage, the adequacy of input recommendation and the degree of application control. The enabling technologies of precision agriculture can be grouped in to fine major categories: Computers, Global Positioning System (GPS), Geographic Information System (GIS), Remote Sensing (RS) and Application control.

Aspects of precision agriculture encompass a broad array of topics including variability of the soil resource base, weather, plant genetics, crop diversity, machinery performance and most physical, chemical and biological inputs used in crop production. Precision agriculture must fit the needs and capabilities of the farmer and must be profitable.

CONTENT ISSUES AND THE GATEWAY DEBATE

Despite all the natural advantages, India's productivity of food grains per hectare is no more than three-fourths of the world average and less than half of that in agriculturally advanced countries, per capita food grain availability even after the Green Revolution, has been less than two thirds of the world average. Only five states in India, namely Himachal Pradesh, Punjab, Haryana, Uttar Pradesh and Madhya Pradesh – produce more grain than their populations can consume. The combined population of the five states is less than one-third of the total of the country. More than two thirds of the population lives in states that are still food-deficit. This requires transport of lakhs of tonnes of food grain, involving high costs and pilferage. Our effort should have been to make all the states self-sufficient with respect to food grains and if some disturbances occurred due to unnoticed natural calamities the nation must be in an ever ready position to mitigate such challenging tasks.

The Indian green revolution is also associated with negative ecological/ environmental consequences. The status of Indian environment shows that, in India, about 182 million ha of the country's total geographical area of 328.7 million ha is affected by land degradation of this 141.33 million ha are due to water erosion, 11.50 million ha due to wind erosion and 12.63 and 13.24 million ha due to water logging and chemical deterioration (salinization and loss of nutrients) respectively. On the other end, India shares 17 per cent of world's population with 2.5 per cent of geographical area, 1 per cent of gross world product, 4 per cent of world carbon emission and hardly 2 per cent of world forest area. The Indian status on environment is though not alarming when compared to developed countries, it gives an early warning to take appropriate precautionary measures. The growth rate of grain production during the Ninth plan has been less than the population growth rate. The poor agricultural performance has not been because of the vagaries of monsoon. From 1997-98 to 2002-03, rainfall was between 92 and 106 per cent of the normal. Per capita availability of grain and per capita calorie intake, which were less than the minimum required for adequate nutrition, have further declined. According to Human Development Report 2003, the percentage of the undernourished in India, which was 21 a few years ago, has now reached 24.

The Government claims that India has emerged as the seventh largest exporter of food grains in the world. This is nothing to be proud of, if we take into account that the total Indian grain export in 2002-03 did not add up to even 4 per cent of the total world exports; and the value of our grain exports did not add up to even the value of our imports of vegetable oils and pulses. The more crucial question, however, is whether it is morally justifiable to export grain when 24 per cent of the population remains under-nourished. The decline in agricultural growth and increase in rural poverty have been due to the long persisting government indifference towards the farm sector, which is evident from plan outlays on Agricultural and Allied activities, Rural Development and Irrigation, which added up to 37.1 per cent of the total during the first plan were brought down to only 19.4 per cent during the Ninth plan. However, the main reason, for poor performance of the farm sector has been the long persisting adverse terms of trade policies for agriculturists in addition to the mismanagement of natural resources leads to ever ending crisis.

THE NEED FOR PRECISION FARMING

The 'Green revolution' of 1960's has made our country self sufficient in food production. In 1947, the country produced a little over six million tonnes of wheat, in 1999; our farmers harvested over 72 million tonnes, taking the country to the second position in wheat production in the world. The production of food grains in five decades, has increased more than three fold, the yield during this period has increased more than two folds. All this has been possible due to high input application, like increase in fertilization, irrigation, pesticides, higher use of HYV's, increase in cropping intensity and increase in mechanization of agriculture.

FATIGUE OF GREEN REVOLUTION

Green revolution of course contributed a lot. However, even with the spectacular growth in the agriculture, the productivity levels of many major crops are for below than expectation. We have not achieved even the lowest level of potential productivity of Indian high yielding varieties, whereas the worlds highest productive country have crop yield levels significantly higher than the upper limit of the potential of Indian HYV's. Even the crop yields of India's agriculturally rich state like Punjab is far below than the average yield of many high productive countries.

NATURAL RESOURCE DEGRADATION

The green revolution is also associated with negative ecological/ environmental consequences. The status of Indian environment shows that, in India, about 182 million ha of the country's total geographical area of 328.7 million ha is affected by land degradation of this 141.33 million ha are due to water erosion, 11.50 million ha due to wind erosion and 12.63 and 13.24 million ha are due to water logging and chemical deterioration (salinisation and loss of nutrients) respectively. On the other end, India shares 17 per cent of world's population, 1 per cent of gross world product, 4 per cent of world carbon emission, 3.6 per cent of CO2 emission intensity and 2 per cent of world forest area. One of the major reasons for this status of environment is the population growth of 2.2 per cent in 1970 – 2000. The Indian status on environment is, though not alarming when compared to developed countries, gives an early warning.

In this context, there is a need to convert this green revolution into an evergreen revolution, which will be triggered by farming systems approach that can help to produce more from the available land, water and labour resources, without either ecological or social harm. Since precision farming, proposes to prescribe tailor made management practices, it can help to serve this purpose.

OPERATIONALIZATION OF PRECISION FARMING

"Agriculture is the backbone of the Indian Economy"-said Mahatma Gandhi five decades ago. Even today, as we enter the new millennium, the situation is still the same, with almost the entire economy being sustained by agriculture, which is the mainstay of the villages. Not only the economy, but also every one of us looks up to agriculture for our sustenance too. Therefore, it is no surprise if agriculture gets the celebrity status in the name of Precision farming (PF).

In recent times, the researchers in the field have been busy formulating methodologies and fabricating new implements for precision farming. It is here the challenge arises considering the implementation of the technology at various levels in the Global community. The need of the hour is not application of the technology but the adoption of appropriate technology, which would suit the particular level of the global community. In India, the farming practices are too haphazard and non-scientific and hence need some forethought before implementing any new technology. It is here that this chapter comes out with a cost-benefit analysis to prove that precision farming is possible in India.

Applications of agricultural inputs at uniform rates across the field without due regard to in-field variations in soil fertility and crop conditions does not yield desirable results in terms of crop yield. The management of in-field variability in soil fertility and crop conditions for improving the crop production and minimizing the environmental impact is the crux of precision farming. Thus, the information on spatial variability in soil fertility status and crop conditions is a pre-requisite for adoption of precision farming.

Space technology including global positioning system (GPS) and GIS holds good promise in deriving information on soil attributes and crop yield, and allows monitoring seasonally-variable soil and crop characteristics, namely soil moisture, crop phenology, growth, evapotranspiration, nutrient deficiency, crop disease, and weed and insect infestation, which, in turn, help in optimizing inputs and maximizing crop yield and income.

Though widely adopted in developed countries, the adoption of precision farming in India is yet to take a firm ground primarily due to its unique pattern of land holdings, poor infrastructure, lack of farmers' inclination to take risk, socio-economic and demographic conditions. The aim of this chapter is to suggest measures for the implementation of this novel technique in the country with a greater emphasis on the systematic approach towards its operationalisation.

Objectives

- To explain the feasibility of precision farming technology with emphasis on seed spacing, tillage, etc.
- To set up a DGPS network all around the country and achieving few centimeters accuracy for the purpose of Site-Specific Management (SSM) in Precision Farming.
- To analyse the cost and benefit in terms of Indian farmer's income-expenditure.

Precision Farming

The conventional agronomic practices follow a standard management option for a large area irrespective of the variability occurring within and among the field. For decades now, the farmers have been applying fertilizers based on recommendations emanating from research and field trials under specific agro-climatic conditions. Since soil-nutrient, characteristics vary not only from one region to another, but also from field to field. Even within a field, there is a need to take into account such variability while applying fertilizers to a particular crop. Consideration of in-field variations in soil fertility and crop conditions and matching the agricultural inputs like seed, fertilizer, irrigation, insecticide, pesticide, etc. in order to optimize the input or maximizing the crop yield from a given quantum of input, is referred to as precision farming or precision agriculture or precision crop management. The term "precision farming" means carefully tailoring the soil and crop management to fit the different conditions found in each field. It is defined as the application of technologies and principles to manage spatial and temporal variability associated with all aspects of agricultural production.. It is also referred to as "prescription farming", "site specific farming" or "variable rate technology."

By catering to this variability, called precision farming, one can improve the productivity or reduce the cost of production and diminish the chance of environmental degradation caused by excess use of inputs. Thus, mapping and analysis of within field variability is an essential input for precision crop management. Thus, PF involves acquiring the variations in crop or soil properties, mapping, and analysing the variations, adopting suitable management techniques to maximize the yield. Farmers have been applying fertilizers based on recommendations emanating from research and field trials under specific agro-climatic conditions, which have been extrapolated to a regional level. Since soil nutrient characteristics vary not only between regions and between farms but also from plot to plot, and within a field or plot, there is a need to take into account such variability while applying fertilizers to a particular crop.

Consideration of in-field/plot variations in soil fertility and crop conditions and matching the agricultural inputs like seed, fertilizer, irrigation, insecticide, pesticide, etc. in order to optimize the input or maximizing the crop yield from a given quantum of input, is referred to as precision farming or precision agriculture or precision crop management.

The information for variability map can be obtained from soil tests for nutrient availability, yield monitors for crop yield, soil samples for organic matter content, information in soil maps, or ground conductivity meters for soil moisture. Generally, the fields are manually sampled along a regular grid and the analysed results of the samples are interpolated using geostatistical techniques.

These techniques are time consuming, labour intensive and in many cases destructive especially, for agricultural situation in India. With small size of landholdings and low income of farmers, the adoption of this methodology in its present form is not feasible. Various workers have shown the advantages of using remote sensing technology to obtain spatially and temporally variable information for precision farming. In an earlier work, Ray et al. (2001) have shown the usefulness of IRS merged data in mapping the variability.

MISCONCEPTIONS ABOUT PRECISION AGRICULTURE

There are several mistaken preconceptions about precision agriculture.

- Precision agriculture is a cropping rather than an agricultural concept. This is due to cropping systems, in particular broad-acre cropping, being the face and driving force of PA technology. However precision farming concepts are applicable to all agricultural sectors from animals to fisheries to forestry. In fact it might be argued that precision farming concepts are more advanced in the dairy industry where the "site" becomes an individual animal, which is recorded, traced and fed individually to optimize production. These industries are just as concerned with improved productivity and quality decreased environmental impact and better risk management as the cropping industry however precision farming concepts have yet to be applied on the same scale in these areas. For example a grazer's use of advance warning meteorological data and market predictions to estimate fodder reserves and plan livestock numbers is a form of precision farming.
- Precision agriculture in cropping equals yield mapping. Yield mapping is a crucial step and the wealth of information farmers are able to obtain from a yield map makes them very valuable. However they are only a stepping-stone in a precision farming management system. The bigger agronomic hurdle lies in retrieving the information in the yield map and using it to improve the production system. The advance

of PA adoption (usefulness) in this country is may soon be bottlenecked at this point due to the lack of decision support systems (DSS) to help agronomists and farmers understand their yield maps. Yield maps may not tell the whole story either with other data sources, *e.g.* crop quality and soil maps, economic indicators or weather predictions, proving further information necessary for correct agronomic interpretations.

- Precision agriculture equals sustainable agriculture. Precision agriculture is a tool to make agriculture more sustainable however it is not the total answer. Precision farming aims at maximum production efficiency with minimum environmental impact. Currently it is the potential for improved productivity (and profitability) that is driving precision farming rather than the more serious issue of long-term sustainability. Precision farming will not fix problems such as erosion and salinity by itself although it will help to reduce the risk of these problems occurring. Sensible sustainable practices still need to be used in conjunction with precision farming.

Obstacles

There are many obstacles to adoption of precision farming in developing countries in general and India in particular. Some are common to those in other regions but the others are specific to Indian conditions are as follows.

- Culture and perceptions of the users
- Small farm size
- Lack of success stories
- Heterogeneity of cropping systems and market imperfections
- Land ownership, infrastructure and institutional constraints
- Lack of local technical expertise
- Knowledge and technical gaps
- Data availability, quality and costs.

Opportunities

Despite the many obstacles listed earlier, business opportunities for precision farming technologies including GIS, GPS, RS and yield monitor systems are immense in many developing countries. The scope for funding new hardware, software and consulting industries related to precision agriculture is gradually widening. In Japan, the market in the next 5 years is estimated at about US $ 100 billion for GIS and about US $ 50 billion for GPS and ₹ Punjab and Haryana states in India, where farm mechanization is more common than in others, may be the first to adopt precision farming on a large scale. Recently, the governments of certain Asian countries initiated special efforts to promote precision farming. In Japan, the Ministry of Agriculture has allocated special funds for research on remote sensing applications of precision farming. A quasi-governmental institute "Bio-oriented Technology Research Advancement Institute (BRAIN)" is also funding research on precision farming. In Malaysia, the Malaysian Agricultural Research and Development Institute (MARDI) is promoting research on precision farming of upland rice. In other countries, the private sector, which holds or leases a large acreage, is likely to adopt precision farming sooner than the small holders.

Precision farming is useful in many situations in developing countries. Rice, wheat, sugar beet, onion, potato and cotton among the field crops and apple, grape, tea, coffee and oil palm among horticultural crops are perhaps the most relevant. Some have a very high value per acre, making excellent cases for site-specific management. For all these crops, yield mapping is the first step to determine the precise locations of the highest and lowest yield areas of the field. Researchers at Kyoto University recently developed a two-row rice harvester for determining yields on a micro plot basis.

Precision farming can bring several benefits to the sugar beet industry in Hokkaido, where the marketing system was changed in 1986 from a quantity (fresh weight) to quality (sugar yield) basis. Heavy N fertilization, a common practice here, results in excessive N levels and a decreased sugar concentration. It is now possible to estimate sugar and amide N concentration in leaves using reflectance in visible bands and root yield using reflectance in visible and infrared bands. Incorporation of such data into a GIS along with precise positioning of non-uniform areas using GPS can be used to vary fertilizer dose within a field, thereby improving productivity. Although weighing conveyor technology has been known for some time, effective yield measurement still remains the main problem in crops such as sugar beet, onion and potato. Further, the preparation of product quality maps for these crops is as important as yield maps. In India, a few researchers in the private sector initiated studies on precision agriculture in high value crops like cotton, coffee and tea. In cotton, remote sensing coupled with GIS can assist in improved precision of insect pest management and harvesting. In Sri Lanka, researchers at the Tea Research Institute are examining precision management of soil organic carbon.

In so far as dairy farming in Asia is concerned, precision farming techniques can help in improving efficiency of methods, timing and rate of application of animal wastes leading to high application efficiency and low environmental pollution. While considering soil and climatic conditions. For instance, factors determining the risk of NO3 leaching, release of N2O through denitrification and contamination of surface and ground water by run-off can be mapped and analysed. Likewise, poorly managed areas in grass lands can be identified and the optimum period for cutting on a plot basis determined.

Nutrient stress management is another area where precision farming can help Indian farmers. Most cultivated soils in India are acidic and spatial variation in pH is high. Detecting nutrient stresses using remote sensing and combining data in a GIS can help in site-specific applications of fertilizers and soil amendments such as lime, manure, compost, gypsum and sulphur. This in turn would increase fertilizer use efficiency and reduce nutrient losses. In semi arid and arid tropics, precision technologies can help growers in scheduling irrigation more profitably by varying the timing, amounts and placement of water. For example, drip irrigation, coupled with information from remotely sensed stress conditions (*e.g.*, canopy temperature) can increase the effective use of applied water from 60 to 95 per cent there by, reducing run-off from 23 to 1 per cent and deep percolation from 18 to 4 per cent.

Pests and diseases cause huge losses to Indian crops. If remote sensing can help in detecting small problem areas caused by pathogens, timing of applications of fungicides can be optimized. Recent studies in Japan show that pre-visual crop stress or incipient crop damage can be detected using radio-controlled aircraft and near-infrared narrow-band sensors. Like wise, airborne video data and GIS have been shown to effectively detect and map black fly infestations in citrus orchards, making it possible to achieve precision in pest control. Perennial weeds, which are usually position-specific and grow in concentrated areas, are also a major problem in developing countries. Remote sensing combined with GIS and GPS can help in site-specific weed management. Although through cost-benefit analysis has not been done yet, the possible use of precision technologies in managing the environmental side effects of farming and reducing pollution is appealing.

Strategies

Precision farming is still only a concept in many developing countries and strategic support from the public and private sectors is essential to promote its rapid adoption. Successful adoption, however, comprises at least three phases including exploration, analysis and execution. Data on crop yield, soil variables, weather and other characteristics are collected and mapped in the exploratory stage, which is important for increasing the awareness among farmers of long term benefits. The approaches to data collection and mapping must, therefore, reflect local needs and resources.

In the analysis stage, factors limiting the potential yield in various areas within a field and their interrelationships are examined using GIS-based statistical modelling. Sadler et al. (1998) showed that quantitatively important yield variation may occur over distances as short as 10m, however, only some factors such as soil structure, water status, pH, nutrient levels, weeds, pests and diseases can be controlled but not the others (soil texture, weather, topography). After determining the significance of each source of variability to profitability of a particular crop and relative importance of each controllable factor, management actions can be prioritized. It must be remembered that in some low yielding areas, the reason for poor yields may be the lack of sufficient soil nutrients in the first place. In such cases, application beyond just replenishment is necessary.

Lastly, execution phase includes variable application of inputs or cultural operations. In most developing countries of Asia. However, it is not always necessary and/or possible to use variable rate applicators. Efforts must, therefore, initially focus on limiting indiscriminate use of inputs in conventional methods. Once the economic and environmental benefits are known widely, variable rate technology would be rapidly implemented at least in high value crops. To spur adoption of precision farming methods in developing countries, pilot demonstration projects must be conducted at various growers locations by involving farmers in all stages of the project.

The pilot projects must attempt to answer the grower's needs and emphasize the operational implementation of technology and complete analysis of the costs and savings involved. Documentation of pilot projects would help in examining the operational weaknesses and identification of remedial measures. The projects can be used to train innovative farmers and early adopters, expose the neighbouring non-participating farmers to the new technologies, and show the usefulness of the technology for short and long-term management.

The role of agricultural input suppliers, extension advisors and consultants in the spread of these technologies is vital. For instance, public agencies should consider supplying free data such as remotely sensed imagery to the universities and research institutes involved in precision farming research. Also, professional societies of agronomy, agricultural informatics, and engineering must provide training guidance in the use of technologies. The involvement of inter/disciplinary teams is essential in this. Small farm size will not be a major constraint, if the technologies are available through consulting, custom and rental services. The role of agricultural cooperatives is important in dissemination of precision farming technologies to small farmers. If precision farming is considered a series of discrete services: map generation, targeted scouting, it is possible to fit these services within the structure of a progressive agricultural cooperative in each developing country.

Changes in agricultural policies are also necessary to promote the adoption of precision farming. There are basically two policy approaches: regulatory policies and market based policies. The former refer to environmental regulations on the use of farm inputs and later refer to taxes and financial incentives aimed at encouraging growers to efficiently use farm inputs. In most developing countries the lack of penalties for pollutant generation has partly contributed to an excessive use of inputs.

Subsidies on inputs and outputs and mechanisms that prevent the price system from rationing limited resources are also common. The latter include state-guaranteed crop prices, tariffs, import quotas, export subsidies. Inputs such as water and fossil fuels are usually sold at prices that are well below the real resource cost of their use, which consists not only production costs but also includes scarcity value and costs of pollution. In such cases, the formulation of policies that reflect the real scarcity value of natural resources and penalize pollution and policies such as green payments for farmers adopting techniques that would lower environmental costs can promote the adoption of precision farming technologies.

In most developing countries, the pollution effects of agriculture have been largely ignored so far because of inability to effectively monitor such effects. The advent of precision farming, and the computerization of input and output flows, will now enable such monitoring. Higher taxes on pollution farms are often recommended, but there is strong opposition to the implementation of the polluter-pays-principle concept in most countries including India. At the same time, some consumers in India would like to see a drastic reduction in the use of pesticides and fertilizers, and are willing to pay as much as 4 to 6 times the normal price for produce such as organic vegetables, soybean and wheat. When the price elasticity of input use is low and the input costs are only a small part of the total production expenditure, as in the case of fertilizers and pesticides. Very high taxes are required to reduce their use adequately. Given the unfeasibility of such high taxes, a hybrid policy may be implemented for controlling pollution. A tax-free quota of N can be combined with taxes on additional N use.

At the research level, many issues remain to be resolved. Although some progress has been made at Space Application Centre, Ahmedabad, yield monitors for small farm conditions are yet to be developed. The development of standards for the hardware and software (image transfer formats and GPS transfer formats, map projection formats) is another issue. Crop models and decision support systems must be improved by considering local resources. Data for calibration of models must be made available to increase their accuracy and/or predictability. The ability to finance a creative information venture in agriculture will affect the speed of diffusion of precision farming technologies. Commercial banks, as well as other sources of funding, have to be educated regarding the potential of precision farming. In many developing countries, it may be worthwhile to develop programmes of subsidized credit to enable R&D activities on precision farming.

PRECISION FARMING IN INDIAN AGRICULTURAL SCENARIO

Agriculture is a very important sector for the sustained growth of the Indian economy. About 70 per cent of the rural households and 8 per cent of urban households are still principally dependent on agriculture for employment. Since some three-quarters of the population live in rural areas, a majority of households thus depend principally on this sector. Though, industrialization of the Indian economy has adversely affected the share of agriculture in the GDP, the fact cannot be ignored that India has undergone a series of successful agricultural revolutions-starting with the 'green' revolution in wheat and rice in the 1960's and 1970's, the 'white' revolution in milk to the 'yellow' revolution in oilseeds in 1980's.

As a result, India has achieved self-sufficiency in agriculture. Applications of agricultural inputs at uniform rates across the field without due regard to in-field variations in soil fertility and crop conditions does not yield desirable results in terms of crop yield.

The management of in-field variability in soil fertility and crop conditions for improving the crop production and minimizing the environmental impact is the crux of precision farming. Geographically, India is widely distributed into several agro-climatic zones, and the information need for the farming systems in these areas is entirely different. Integrating the application of available technologies to realize farmers' goals requires a systems approach to farming. The concept of fully automated villages was a dream till few months back. But the reality has come to true with the rapid growth of information and communication technology in the world scenario. The wired villages and info villages have shown that Information can be disseminated in more useful manner and as farmers need.

PRECISION AGRICULTURE

Precision agriculture is an agricultural system that has the potential of dramatically changing agriculture in this 21st century. Precision agriculture lends it self to most agricultural applications and can be implemented at whatever levels are required. Precision agriculture is based on information technology, which enables the producer to collect information and data for better decision making. Precision agriculture is a pro-active approach that reduces some of the risk and variables common to agriculture. Precision agriculture is more environmentally sound and is and integral part in sustaining natural resources. To better understand the need for an accurate definition of precision agriculture lets look at how precision agriculture is being considered. Precision agriculture is considered a concept, management strategy, and even a philosophy. It is said, "Precision agriculture is a phrase that captures the imagination of many concerned with the production of food, feed, and fibre." The concept of precision agriculture offers the promise of increasing productivity while decreasing production cost and minimizing environmental impacts. Precision agriculture conjures up images of farmers overcoming the elements with computerized machinery that is precisely controlled via satellites and local sensors and using planning software that accurately predicts crop development. This image has been called the future of agriculture. Precision farming is characterised by a number of sophisticated tools that assist in monitoring variation and managing inputs. These include:

- Global Positioning System (GPS) – a referencing device capable of identifying sites within a field;
- Sensors and dataloggers – crop, soil and climate information can be monitored at a high frequency using these technologies;
- Geographic Information Systems (GIS) – maps of these attributes can be generated and analysed using simple browsers or complex models.

GPS ROLE IN PRECISION AGRICULTURE

Precision Agriculture is doing the right thing, at the right place, at the right time. Knowing the right thing to do may involve all kinds of high tech equipments and fancy statistics or other analysis. Doing the right thing however starts with good managers and good operators doing a good job of using common tools such as planters, fertilizer applicators, harvesters and whatever else might be needed..

The use of GPS in Agriculture is limited but it is fair to expect wide spread use of GPS in future. Recently a GPS-based crop duster (precision GPS Helicopter), which can spray an area as small as 4 X 4 mtr. is attracting great attention. Some progressive farmers are now beginning to use GPS for recording observations. Such as weed growth, unusual plant stress, colouring and growth conditions, which can then be mapped with a GIS programmes. In the years to come, GPS system role in precision agriculture may help the Indian farmers to harvest the fruits of frontier technologies without compromising the quality of land and produce.

STATUS OF PRECISION FARMING IN INDIA

Hence, the first thing that comes to mind is that, this system is not for developing countries, especially India, where the farmers are poor, farming is mostly subsistent and the land holding size is small. But, this is far from the truth as this approach has a large potential for improving the agricultural production in developing world.

Imagine this situation where a farmer goes to his field with a GPS (Global Positioning System)-guided tractor. The GPS senses the exact location of tractor within the field. It sends signals to the computer fixed on to the tractor, which has a Geographical Information System (GIS), storing the soil nutrient requirement map in it. The GIS, in consultation with a Decision Support System would decide what is the exact requirement of fertilizers for that location. It then commands a variable rate fertilizer applicator, which is again attached with the tractor, to apply the exact dosage at the precise location of farm.

But, this is what precision farming means to large growers in the highly developed parts of the globe. To make it clearer, Precision Farming is the system of matching of resource application and agronomic practices with soil attributes and crop requirements since they vary across a field.

Tata Kisan Kendra: The concept of precision farming being implemented by the TKKs has the potential to catapult rural India from the bullock-cart age into the new era of satellites and IT. TCL's extension services, brought to farmers through the TKKs, use remote-sensing technology to analyse soil, inform about crop health, pest attacks and coverage of various crops predicting the final output. This helps farmers adapt quickly to changing conditions. The result: healthier crops, higher yields and enhanced incomes for farmers.

Government organization: Precision Agriculture models are not complete, unless the parameters related to empowerment of the farmers; especially small and marginal farmers are integrated. Now it is the turn of good news to the Indian farming community.

Some of the research institutes such as Space Applications Centre (ISRO), M.S. Swaminathan Research Foundation, Chennai, Indian Agricultural Research Institute, New Delhi, and Project Directorate of Cropping Systems Research, Modipuram, had started working in this direction and in soon it will help the Indian farmers harvest the fruits of frontier technologies without compromising on the quality of land.

According to the Exim Bank officials, though the research and development on PF is currently at a nascent stage in the country, the efforts being put on by the four research institutes were expected to turn the green revolution into an evergreen revolution.

In this context, ISRO has also initiated Gramsat project in Orissa. In the line of JDCP, the Gramsat project aims at empowering the people especially the poor and marginalized, by awareness building and access to information and services. Towards this, a network of one-way video and two-way audio Forecasting the yield of mono and multiple crops is being done at NRSA.

Acreage estimates and crop inventory is being done during Kharif and Rabi seasons for Rice, which is the major crop grown in our India. Other crops like Banana, Chillies, Cotton, Maize, Sugarcane and Tobacco are also being inventoried. Satellite data can also delineate different crops that are grownin the same area, and an inventory of each of the crops can be done.

Remote Sensing and Sensors for PF

Precision farming needs information about mean characteristics of small, relatively homogeneous management zones. This is for Data acquisition of the farms to find the soil, vegetation and other parameters that are amenable for remote sensing. Remote sensing techniques play an important role in precision farming by providing continuous acquired data of agricultural crops. Remote sensors image vegetation, which is growing on different soil types with different water availability, substrate, impact of cultivation, and relief Sensors use for the following applications:

- Soil Properties Sensing: Soil Texture, Structure, and Physical Condition Soil Moisture; Soil Nutrients.
- Crop Sensing: Plant Population; Crop Stress and Nutrient Status.
- Yield Monitoring Systems: Crop Yield; Harvest Swath Width; Crop Moisture:
- Variable Rate Technology Systems: Fertilizer flow; Weed detection, pressure sensors

PRACTICAL PROBLEMS IN INDIAN AGRICULTURE

Precision Agriculture has been mostly confined to developed countries. Reasons of limitations of its implementation in developing countries like India are:

- Small land holdings,
- Heterogeneity of cropping systems and market imperfections,
- Lack of technical expertise knowledge and technology (India spends only 0.3 per cent of its agricultural Gross Domestic Product in Research and Development)
- High cost.

In India, major problem is the small field size. More than 58 per cent of operational holdings in the country have size less than 1ha. Only in the states of Punjab, Rajasthan, Haryana and Gujarat more than 20 per cent of agricultural lands have operational holding size of more than four hectare.

There is a scope of implementing precision Agriculture for crops like, rice and wheat especially in the states of Punjab and Haryana. Commercial as well as Horticultural crops shows a wider scope for precision Agriculture.

In India, broadly two types of agriculture *viz.*, high input agriculture characterized by the provision of assured irrigation and other agricultural inputs, and subsistence farming, which is confined mostly to rain-fed, or dry land regions, are prevalent. Nearly two-third arable land in India is rain-fed. The crop yields are very low (˜ 1 t ha-1) and very good potential exists for increasing productivity of rain-fed Cropping systems.

METHODOLOGY

Using a GPS along with a digital drainage map, the farmer is able to apply these pesticides in a safer manner. The spraying equipment can be preprogrammed automatically to turn off when it reaches the distance limitation or zone of the drainage feature. Additionally, farmers can preprogram the rate of pesticide or fertilizer to be applied so that only the amount needed determined by the soil condition is applied varying this rate from one area of the field to another. This saves money and allows for safer use of these materials. Data collected by the GPS operations can be automatically recorded with the GIS programme. Remotely sensed data can be analysed and added to the GIS using soil maps, digital terrain and field operations information as ground truth. Additionally the farmers are able to record observations through the growing season such as weed growth, unusual plant stress or coloring and growth conditions. This can be used to guide further field operations like spraying, fertilizing and irrigating and it is part of the permanent record. Precision farming will make a strong impact on the way farmers manage their farm operations in the future.

Seed Spacing

The yield from the seeds is maximum when adequate care is taken to adequately space them to enable each plant the optimum requisites for its growth, namely the soil nutrients, water, sunlight and protection against pest infestation.

This is best accomplished by varying the seeding rate even within a single piece of land according to the soil conditions such as texture, organic matter and available soil moisture.

One would plant fewer seeds in sandy soil as compared to silt loam soils because of less available moisture. The lower seed population usually has larger heads (ears) of harvested seeds providing for a maximum yield. Since soils vary even across an individual farm field, the ability to change seeding rates as one goes across the field allows the farmer to maximize this seeding rate according to the soil conditions. A computerized soil map of a specific field on a computer fitted on the tractor along with a GPS can tell farmers where they are in the field allowing the opportunity to adjust this seeding rate as they go across their fields.

STEPS TO BE TAKEN FOR IMPLEMENTING PF

In the present existent situation, the potential of precision agriculture in India is limited by the lack of appropriate measurement and analysis techniques for agronomically important factors. High accuracy sensing and data management tools must be developed and validated to support both research and production. The limitation in data quality/availability has become a major obstacle in the demonstration and adoption of the precision technologies. The adoption of precision agriculture needs combined efforts on behalf of scientists, farmers and the government. The following methodology could be adopted in order to operationalise precision farming in the country.

- Creation of multidisciplinary teams involving agricultural scientists in various fields, engineers, manufacturers and economists to study the overall scope of precision agriculture.

- Formation of farmer's co-operatives since many of the precision agriculture tools are costly (GIS, GPS, RS, etc.).
- Government legislation restraining farmers using indiscriminate farm inputs and thereby causing ecological/environmental imbalance would induce the farmer to go for alternative approach.
- Pilot study should be conducted on farmer's field to show the results of precision agriculture implementation.
- Creating awareness amongst farmers about consequences of applying imbalanced doses of farm inputs like irrigation, fertilizers, insecticides and pesticides.

Realizing the potential of space technology in precision farming, the Department of Space, Government of India has initiated eight pilot studies in well-managed agricultural farms of the ICRISAT, the Indian Council of Agricultural Research and the Agricultural Universities, as well as in farmers' fields. The pilot studies aim at delineating homogeneous zones with respect to soil fertility and crop yield, estimation of potential yield, yield gap analysis, monitoring seasonally-variable soil and crop conditions using optical and microwave sensor data, and matching the farm inputs to bridge the gap between potential and actual yield through Spatial Decision Support Systems (SDSS). The test sites are spread over a fairly large area across a cross section of agro-climatic zones of the Indian sub-continent, and cover some of the important crops like wheat, rice, sorghum, pigeon pea, chickpea, soybean and groundnut.

The next step would be to generate detailed-level information on soil resources addressing potentials and limitations of individual fields since except for states like Punjab, Haryana, Madhya Pradesh and Maharashtra where fields size is quite large, practically individual field could be treated as a homogenous management unit for the purpose of precision farming.

7

Chemical Composition of Membranes Structure

CHEMICAL COMPOSITION OF MEMBRANE

Chemical membranes are composed primarily of two main classes of compounds, lipids and proteins, which interact in several ways with water to bring about a characteristic trilaminar structure. As a very broad generalization it might be said that lipids make up the supporting matrix of the membrane while the proteins thus supported determine its specific properties. This is, of course, merely a convenient simplification as it is becoming plain that some of the characteristic properties of a membrane, particularly those which determine its effectiveness as a barrier to the diffusion of water and electrolytes, depend on the nature of the membrane lipids. Conversely some proteins have a structural role. For the present purposes, however, the classes of compounds will be considered separately and a synthesis attempted later on.

LIPIDS

Of the lipids present in plant cells the various phospholipids, glycolipids and sterols are of the greatest significance in membrane construction. The relative abundance of the components can be quite variable depending on the part of the plant analysed, the species and prevailing environmental conditions.

Phospholipids

The phospholipid molecule can be separated into a charged or polar 'head' region and an uncharged or non-polar 'tail'. Such a molecule is described as amphipathic, and as we shall see later on, this property is of crucial importance in determining membrane structure.

Phospholipids are generally thought to be restricted to membranes but the extremely rapid rate at which membranes can be taken apart and re-assembled, as in cell plasmolysis and de-plasmolysis, makes it probable that there are stores of phospholipid within the cell.

Phospholipids are readily extracted from macerated plant tissues by a mixture of chloroform and methanol (2:1) and can be separated by thin layer chromatography using a variety of solvent systems.

The commonest phospholipids in plant membranes are derivatives of phosphatidic acid; thus, lecithin is the choline ester of phosphatidic acid. Phosphatidic acid (PA) itself is generally said to occur only in minute quantities in membranes or not at all, indeed, its presence in an extract is often taken as an indication of the activity of phospholipase D (Mazliak, 1973). There is a report, however, in which phosphatidic acid is said to be one of the major constituents of the plasmalemma of oat (*Avena sativa*) root, (Keenan *et al.,* 1974). Unfortunately, detailed analyses of the plasmalemma from other plants are not available for comparison. In passing it might be noted that a great deal remains to be done, firstly in preparing pure sub-cellular fractions of the plasmalemma and of other membranes from plants and subsequently in determining their lipid composition. Some of the available information on the distribution of the different types of phospholipid.

The information on the composition of the membranes of mitochondria and chloroplasts is the most detailed and reliable since these organelles can be separated with relative ease and high purity during cell fractionation. The predominant phospholipid in a given membrane may be characteristic, *e.g.* phosphatidylglycerol (PG) is a major component of chloroplast membranes while it is only a minor component of the inner mitochondrial membrane where diphosphatidylglycerol (DPG) is predominant. In general extracts of shoots and roots neither of these phospholipids is as abundant as lecithin (PC) or phosphatidyl ethanolamine (PE). A small quantity of phosphatidyl inositol (PI), usually less than 10 per cent of the total phospholipid, is found in all membranes.

The exact chain length of the acyl groups R_1 and R_2 which make up the hydrophobic tail, is not defined precisely. In nature it can vary considerably even in one type of phospholipid from a given tissue. The chain may be made from 12 to 22 carbon atoms and may contain up to three or, rarely, six double bonds. The chain is straight in all eukaryotic organisms and has been found to be branched only in certain bacteria (Asselineau, 1966). Variation in both the length and unsaturation (*i.e.* the number of double bonds) of the hydrocarbon chain influences its melting point; shorter and unsaturated chains melt at much lower temperatures than longer and saturated ones.

As an example of this consider the effect of double bonds on the melting of free fatty acids containing 18 carbon atoms; the saturated stearic acid ($C_{18:0}$) melts at 69°C, the monounsaturated oleic acid ($C_{18:1}$) at 5°C and the double unsaturated linoleic acid ($C_{18:2}$) at –12°C. Organisms which live in warm conditions and warm blooded animals are generally found to have phospholipids with an abundance of fatty acids which tend to be fully saturated. By contrast, organisms which are exposed to lower temperatures have either more unsaturated acids or ones with shorter average chain lengths or a combination of both of these to give phospholipids whose tails remain fluid. The significance of the maintenance of membrane fluidity will become apparent later. The process of hardening plants against injury from frost or chilling is accompanied by changes in the degree of unsaturation of the membrane lipids.

The fatty acid analyses of individual phospholipids extracted from various sources. Bearing in mind that there is a great deal of room for manoeuvre in selecting the fatty acids to suit the prevailing environmental temperature the values for the relative abundance of fatty acids should be considered only as very general guides to the types of acid found in nature. Thus, the predominant fatty acids have even numbers of carbon atoms, the saturated acids found most frequently are palmitic (16:0) and stearic (18:0), and the principal unsaturated acids are linoleic (18:2) and the triply unsaturated linolenic (18:3). The fatty acid composition of lecithin can depend very strongly on its origin. For example, the lecithin in the outer mitochondrial membrane is much richer in palmitic acid (16:0), and perhaps is a less fluid component than in the inner mitochondrial membrane where triply unsaturated linolenic (18:3) is the most abundant fatty acid.

Glycolipids

In several respects the glycolipids resemble phospholipids. The molecule is amphipathic, the polar group being a galactosyl derivative of a diglyceride, the non-polar part of the molecule being a pair of long, straight-chain fatty acids. Glycolipids are unusually rich in the triply unsaturated linolenic acid ($C_{18:3}$) which may make up more than 90 per cent of the fatty acid.

The two most abundant glycolipids are mono- and di-galactosyl diglyceride. They are characteristic of photosynthetic tissues since they are the major lipid component of chloroplast lamellae, largely replacing the phospholipids. Ongun *et al.,* (1968) showed that more than 80 per cent of all of the glycolipid in leaf cells was present in the chloroplasts. The probable orientation of these molecules in the chloroplast lamellae is much like that described for phospholipids with the fatty acid tails inserted into the central region of the membrane with the polar galactosyl groups at the membrane surface protruding into the stroma (Weier and Benson, 1967). Because the fatty acid is so highly unsaturated the membranes of lamellae probably remain fluid even at sub-zero temperatures—thus any photosynthetic reaction, or molecular reorientation, which depends on membrane fluidity may have a wide temperature range in which it can occur.

A sulphur-containing glycolipid is found as a minor component of most membranes. It is known as sulpholipid and its structure and occurrence in chloroplasts was reported by Benson (1963); the acyl groups are mainly palmitic with a preponderance of linolenic acid, thus resembling the other galactosyl lipids. Sulpholipid represents only 1 per cent of the total lipid in most tissues and organelles but in chloroplasts it may be as much as 10–15 per cent of the lipid (Ongun and Mudd, 1968).

Sterols

A number of sterols can be extracted from plant tissues and fungi as well as from isolated membrane fractions. The conventional example of a sterol of common biological origin is cholesterol; in practice plant cells contain relatively little of this sterol in comparison with animal cells. Such meagre quantitative data as is available show that sterols having 29 carbon atoms, *e.g.* *b* -sitosterol, are the most abundant in higher plants, while in fungi the C-28 sterol, ergosterol is often dominant. All of these molecules have an extended concertina-like configuration (known as the 'chair' or 'boat'), seem metabolically inert and are synthesized and turned over very slowly, especially in comparison with the other lipid components of membranes (Nes, 1974).

Their function in membranes is not well understood but it is likely that they have an architectural role concerned with the maintenance of structure or order in the lipid domain. In this respect all of them probably function in the same was as cholesterol because Butler *et al.,* (1970) found that the structural order of bilayer membranes synthesized from lipids of ox brain tissue was stabilized equally well by cholesterol, *b* -sitosterol of plant origin and ergosterol.

In the plasmalemma of the animal and plant cell there is a much higher proportion of sterols and sterol esters relative to phospholipid than in other membranes. It should be noted, however, that the membranes of intracellular organelles contain much more protein than do plasmamembranes. To some extent this protein, much of which is bonded hydrophobically to the lipid, may function in a way similar to sterol in maintaining the structural order of the membrane interior.

Proteins

In many membranes, particularly in those of chloroplasts and mitochondria proteins make up most of the weight. The proteins found are many but as a first step in classifying them *integral* and *peripheral* proteins may be distinguished.

This classification anticipates the subsequent discussion of membrane structure on page 38 but the terms clearly suggest that proteins in the two classes are associated with other components in the membrane in different ways. The recognition that certain proteins are embedded deeply in the lipid membrane represents a departure from the view, often advanced in earlier texts, that all of the membrane protein is located in the two peripheral bands which stain darkly with osmium and are visualized in the electron microscope. Whereas some of the protein is certainly located in this way and is probably bonded to the polar regions of the phospholipids electrostatically, it has become apparent that much protein is associated with the non-polar regions of the lipid by hydrophobic bonding.

Peripherally located protein can be easily separated by washing with salt solutions or chelating agents but integral proteins are attached very strongly to the membrane and can be removed only after drastic treatment with organic solvents or detergents; even then, the isolated protein usually has some lipid attached to it.

Most of the membrane-bound enzymes, transport proteins (*e.g.* monovalent cation stimulated ATPase), drug and hormone receptors (in animal cells) and antigenic proteins are *integral* and are revealed when membranes are split open in freeze-fracturing. In many instances the enzymes arenonfunctional *in vitro* in the absence of lipid. It is thought that the non-polar parts of the polypeptide chains are associated with the hydrophobic tails of the fatty acids and, this being so, several types of conformation are possible.

Proteins, like phospholipids, are amphipathic and their polar regions will arrange themselves so that their contacts with the hydrophobic regions of the membrane will be minimized; to ensure this, a protein could be arranged so that its polar, hydrophilic region lies among the phospholipid 'heads', or projects through them into the protein at the membrane periphery.

Long polypeptide chains with charged groups at either end may actually lie across the width of the membrane, or groups of them might lie with polar groups directed inwardly to form a hydrophilic pore across the membrane. An alternative conformation would be provided by the formation of a globular structure in which all of the polar groups would be directed towards the centre of the globule so that a hydrophobic surface would be presented to the lipid. This latter kind of conformation is probably least common.

The peripheral proteins can be attached to the polar groups of either phospholipids or integral proteins; examples which might be taken include cytochrome *c* which is located on the outer surface of the inner mitochondrial membranes, the chromoprotein, phytochrome, which is thought to be attached to the plasmalemma and the sulphate and other ion-binding proteins on the outer surface of bacterial membranes.

From a quantitative point of view, certain generalizations about the relative abundance of peripheral and integral proteins can be made. The greater the metabolic activity which centres upon a given membrane system the greater amount of protein integrated into it.

Thus, it might be anticipated that chloroplast lamellae and inner mitochondrial membranes would be relatively rich in these proteins, whereas membranes whose role is more concerned with providing a diffusion barrier, *e.g.* the plasmalemma and the tonoplast, would be less so; evidence from electron microscopy shows that this is so.

Water

Water is an important, if neglected, constituent of membranes for several reasons. In a general way it determines their basic design since, in its presence, amphipathic lipids assume a bilayered configuration. There are, however, other specific associations of water molecules and membrane components which are not fully understood.

It has been estimated that water of hydration accounts for about 30 per cent of the weight of membranes. Much of this water will almost certainly not be in a liquid state but will exist in ordered layers around the hydrophilic parts of lipids and proteins. Immobilized by hydrogen bonding these water molecules are in a liquid-crystalline condition and cannot be frozen to form ice.

Water layers bound at the surface of the membrane have been estimated to have viscosity of 39 times that of pure water and to have a thickness of at least 2.2 nm (Schultz and Asunmaa, 1970). They must contribute to the mechanical stability of membranes and add significantly to their barrier properties to diffusing solutes. Expressing an extreme view, Ling (1973) has proposed that it is not lipid but these polarized multilayers of water which provide the cell with its selective surface barrier.

Hydrophobic bonding between the non-polar regions of lipids and integral proteins is favoured thermodynamically by the interactions of their polar regions with water (Tait and Franks, 1971). Much experimental evidence points to the fact that water molecules are not restricted to membrane surfaces but cross the hydrophobic regions in numerous water-filled pores. These pores are thought to conduct water and small solutes (diameter <0.4 nm) to which membranes are highly permeable. Some water lining these pores is fully 'organized' and should, therefore, be regarded as a structural feature but there is indirect evidence to suggest that some of it must be free water *in transit.*

TRANSPORT OF SUBSTANCES ACROSS MEMBRANES

In a healthy cell there is a continuous interchange of water, ions, uncharged solutes, metabolites and dissolved gases across the plasmalemma. As one might expect, not all of these substances move through the membrane in the same manner. Firstly, some substances diffuse into a cell down a gradient of potentia energy; such movement is spontaneous and is the thermodynamic equivalent 0^1 heat passing from a warmer to a cooler body.

There are, however, many substances which are accumulated by cells against a gradient of potential hence their movement into the cell is 'uphill' and is equivalent to the flow of heat from a cooler to a warmer body. 'Uphill' transport requires work to be performed and thus consumes energy. 'Downhill' transport is frequently described as passive while 'uphill' transport is described as active and directly involves the participation of cellular metabolism.

Thus, passive movements may occur by at least three types of pathway, whereas active movements must be linked to some energy-consuming mechanism, referred to as a 'pump', in the membrane. The third type of movement occurs because of the undulation and vesicularization of the membrane in endocytosis. Clearly this is a process which depends at some point on metabolism but, as is discussed below, the substances which move into the cell do not necessarily cross the membrane at all. In such circumstances the observed transport is not strictly *active* in a thermodynamic sense.

CRITERIA FOR ACTIVE ('UPHILL') TRANSPORT

To decide whether or not an ion or solute is actively transported across a membrane we need to know its activity or concentration in the two solutions separated by the membrane and for ions, in addition, we must know the electrical potential difference across the membrane. It is frequently difficult to measure the concentrations, and more difficult to measure the activity, of substances within cells with any accuracy, especially in those of higher plants.

In the giant cells of several sorts of algae, which may be 5,000 to 10,000 times the volume of a parenchyma cell in a root, such measurements are made routinely. The electrical potential difference across membranes can be measured if a small glass micro-electrode, with a tip diameter of 1 to 3*m* m, can be inserted into the membrane-bounded compartment.

Having made the necessary measurements a simple test can be applied to see if a given ion or solute within the compartment is at a higher or lower potential than in the surrounding solution. The principal snag in this analysis is that the cell or compartment should be in a steady state and that no net movement of solute should be occurring. In nature this condition is infrequently met.

THE NATURE AND ORIGIN OF THE MEMBRANE POTENTIAL

It is clear that electrical potential differences across membranes are of great importance in generating driving forces on ions. It is important, therefore, to try to understand how these potentials arise and how they are maintained.

An electric potential difference arises because positive and negative charges become separated. Since the cytoplasm of most cells is electrically negative relative to the surroundings it is very slightly enriched in anions relative to cations. This can be attributed to the differential permeability of the cell membrane and to the activity of ion pumps. First let us examine how differential permeability can create an electrical potential difference.

Diffusion Potential

Imagine a simple system of two compartments separated by a membrane which has a much higher permeability to K^+ than to Cl^-. If the compartments are filled with potassium chloride solutions of different concentration, initially K^+ will move through the membrane out of the more concentrated compartment, and for a very brief period, the more concentrated cell will lose K^+ faster than Cl^- leaving it enriched in negative charge.

The negative *diffusion potential* thus created slows down the further escape of K^+ by attracting it back into the more concentrated compartment. When the potential has developed, a large concentration difference can be maintained between the compartments. Since the membrane has a finite permeability to Cl^-, albeit a low one, over a long period of time both the electrical potential difference (the diffusion potential) and the concentration difference would run down as Cl^- leaked through the membrane. If the membrane were completely impermeable to Cl^-, the potential, once established, would be maintained indefinitely. In nature, membrane permeability to anions is a tenth to one hundredth of that for the monovalent cations; thus, the maintenance of a potential of the kind just considered depends on topping up the cell with anions at a rate comparable with their leakage into the surroundings. This is an 'uphill' transport and therefore requires the mediation of some ion pumping mechanism. Active transport is, therefore, necessary to maintain a diffusion potential.

MEMBRANE PUMPS

Pumping mechanisms which allow cells to accumulate solutes up gradients of potential can contribute to the membrane potential in both an indirect and a direct way. The former type are referred to as neutral exchange pumps in which the cell dumps an unwanted ion of equal and like charge into the external environment in a one-to-one exchange for an ion which is more useful, *e.g.* there are well known exchanges of cellular Na^+ or H^+ for K^+ from the surroundings.

The second type transports an ion in one direction only without coupled exchange and is known as *electrogenic* since charge is separated. They can, therefore add to, or subtract from the diffusion potential, described above, depending on which ion is carried.

Neutral Ion Pumps

Neutral ion pumps are present in the plasmalemma of all cells and by their activity they create the ionic asymmetry necessary to set up the diffusion potential described above. As a much simplified illustration of this, consider a cell which, because of its synthetic and respiratory activity, is generating H^+ and HCO_3^- internally. A pair of exchange pumps could swap H^+ and HCO_3^- for K^+ and Cl^- from the surroundings quickly enriching the interior in these ions and thus setting up the conditions in which the differential rates of diffusion of K^+ and Cl^- out of the cell give rise to a membrane potential. But how does such a pump actually work? There are fewer detailed examples than one would like but the best known is of the membrane-bound ATPase which exchanges intracellular Na^+ for extracellular K^+ in many animal and plant cells. In the red blood cell it is known that Na^+ is one of the cofactors which is essential for the binding of ATP to the ATP-ase enzyme. *In vivo* the active centre of the enzyme is accessible only from the cytoplasm, so that the ATP and the Na^+ must be inside the cell.

Once bound, the ATP is hydrolysed, ADP is released into the cytoplasm leaving the cleaved terminal phosphorus atom attached to the active centre to form a phosphoenzyme. These reactions result in some molecular re-orientation of the phosphoenzyme and its attached Na^+ which exposes the ion-binding site to the different chemical environment of the external medium. It is proposed that this change of environment alters the ion-specificity of the binding site so that K^+ is favoured; K^+ thus replaces Na^+. This done, there is a second re-orientation (referred to as 'relaxation') which carries the bound K^+ to the inside. The phosphorus is released from the active centre and Na^+, which is preferentially bound on the cytoplasmic side exchanges for K^+ and the pump is ready for a second cycle. The pump has used the free energy released on hydrolysis of ATP as fuel to exchange K^+ and Na^+ against their respective electrochemical potential gradients. The two ions in the appropriate orientations are essential cofactors in the enzyme reaction; *in vitro,* ATPase of this kind will not hydrolyse ATP unless Na^+ and K^+ are both present.

In theory many pumps based on ATPase are possible with only subtle modifications of the ATPase molecule to provide binding sites of varying field strength which will select various ions, *e.g.* a Ca^{2+} transporting ATPase is found in mitochondria and in sarcoplasmic reticulum (Racker, 1972).

Electrogenic Pumps

The unidirectional transport of an ion across a membrane separates charges and in so doing provides a driving force for the passive diffusion of a similarly charged ion in the opposite direction or an oppositely charged ion in the same direction. The molecular details of exactly how an electrogenic pump is put together remain uncertain although in one instance it is highly likely that an electrogenic H^+ -efflux pump is based on an ATPase (Slayman *et al.,* 1973). It is possible, nevertheless, to deduce certain general consequences of their operation. If, for instance, there was an outwardly directed pump at the plasmalemma which actively pumped hydrogen ions (protons) out of the cell thus making the interior electrically negative, this could contribute to the electrical driving force on the diffusion of K^+ from the external medium. Indeed the rate at which charge is extruded and the rate at which it leaks back into the cell must be very nearly in balance unless a dangerously large potential is to accumulate. Examples of both proton extrusion pumps and anion influx pumps of the electrogenic kind are well documented from research on plant tissues.

In the giant alga, *Acetabularia,* an electrogenic chloride influx pump contributes more than half of the potential of –170mV found across the plasmalemma when the cell is kept in the light. Almost immediately the cell is put in the dark the pump stops working (since it is closely linked with photosynthesis) and the membrane potential abruptly depolarizes to –80mV (Saddler, 1970). A similar light-dependent electrogenic pump is found in *Nitella translucens* (Spanswick, 1972, 1974). Electrogenic pumps are not, however, restricted to green tissues but have been reported in plant roots (Higinbotham *et al.,* 1970) and fungal hyphae (Slayman, 1970). In every instance, however, inhibition of the pump caused an immediate depolarization of the membrane potential, indeed this is often used to detect the activity of such a pump. The inhibition of a neutral ion pump gives rise to gradual depolarization as the ionic asymmetry runs down.

MEMBRANE CARRIERS

Many ions and uncharged solutes cross membranes more rapidly than could be expected if they passed through the membrane lipids without some assistance. Since pores in membranes are too narrow to accommodate many substances which are transported, it is widely believed that membranes contain carrier molecules which, in combination with the solute, facilitate diffusion.

The ion pumps considered above are carriers of a special kind since they are linked to the metabolic activity of the cell; the carriers we shall now consider promote net movements of solutes only down the prevailing potential gradient and are not capable of 'uphill' transport even if in some instances they appear to be doing so.

Evidence from Kinetics

One widely used approach to gather information about carriers has been the application of kinetics first derived from enzyme reactions. The evidence for these carriers is obtained by placing a cell or tissue in a range of solute concentrations and measuring the initial rate of uptake. The uptake rate shows a tendency to saturate at higher concentrations and can thus be used to calculate the maximum velocity, V_{max}, possible under the conditions used in the experiment.

By making a double reciprocal plot of the data the concentration of solute at which half maximal velocity is achieved can be estimated. This is known as the Michaelis Constant, K_m. The K_m measures the affinity of the carrier for the solute it carries; if the affinity is high then the concentration, K_m will be low and *vice versa.* For most ions in plant tissue K_m is quite low, concentrations ranging from 5–100 *m* M, but for sugars and other metabolites K_m values are usually greater than 300 *m* M. Much work of this kind is summarized by Epstein (1972) who shows that at concentrations less than 0.1 mM, the uptake of a given ion is not subject to serious interference from other common ions in solution. There is, however, competitive inhibition between related ions of similar molecular dimensions, *e.g.* K^+ uptake is inhibited competitively by Rb^+ but not by Na^+; Ca^{2+} is inhibited by Sr^{2+} but not by Mg^{2+}. Thus the carriers which bind the major nutrient ions at low concentrations appear to be highly ion-selective. At higher concentrations (more than 1.0–10.0 mM) this selectivity begins to decline. The interpretation of this observation is contentious and beyond the scope of this chapter but can be pursued in Epstein (1972), Laties (1969) and Clarkson (1974).

The limitation of the kinetic approach is that it can tell us nothing about the nature of the carrier. One can observe similar uptake kinetics for ions whose transport into the cell must be mediated by ion pumps *e.g.* $H_2PO_4^-$ and Cl^- as for ions which probably diffuse into the cell passively *e.g.* Na^+ and Ca^{2+} and for those which are completely exotic and toxic, *e.g.* Tl^{4+} (Barber, 1974). Indeed, it has been pointed out that saturation kinetics of this kind would also be found if salt movement was observed across a synthetic membrane containing nothing but pores (Stein and Danielli, 1956), where the system would saturate when all of the pores were filled with solute at any moment in time; V_{max} is, after all, merely a measurement of capacity to react or transport and K_m is derived from it.

Ionophores as Lipophilic Carriers

A more illuminating approach to the nature of carriers has come from studies on the ionic conductance of synthetic membranes which have been modified in various ways. A bilayer of pure phospholipids has a very low conductance to ions, usually only 10^{-7} to 10^{-8} ohm^{-1} cm^{-2}. The addition of very small amounts of ionophores (*i.e.* ion-carrying antibiotics) like monactin or valinomycin to the solutions bathing the synthetic bilayer causes a huge increase in the conductance. 10^{-6} M monactin changes the membrane conductance to K^+ nearly a million-fold and that even at 10^{-10} M its effect is quite strong.

The conductance change for a given monactin concentration is greatest for K^+ and for Rb^+ and is much less for Na^+, Cs^+ and Li^+. Monactin is, therefore, acting as a selective carrier of K^+ and Rb^+ and valinomycin, another bacterial product, behaves similarly. Since these two compounds differ chemically it is instructive to see what they have in common. Both of them are amphipathic ring-structured molecules which have their non-polar groups on the outside of the ring and their polar groups directed towards the space at the centre of the molecule. The outside of the ring interacts favourably with lipid while the hydrophilic core, 0.7 nm in diameter, provides room for several hydrated potassium ions to be bound. Evidence from a variety of sources shows that this complex diffuses across the membrane so that the ions never leave a polar environment (Eisenman *et al.,* 1968).

Other substances are known which select for divalent cations, *e.g.* the unnamed compound A23187 which is a carboxylic acid antibiotic found in cultures of *Streptomyces chartreusensis* (Reed and Lardy, 1972). This compound carries Ca^{2+} and Mg^{2+} across bilayers and natural membranes but has no effect on monovalent cations. Apart from a few synthetic analogues all of the ion-carrying antibiotics are natural products of bacteria and fungi. There are many who believe that compounds of a similar kind may act as ion carriers in all membranes but the technical difficulty of isolating what are probably minute quantities of such compounds from tissues appears to be formidable and so the belief may rest on faith for some time yet.

Co-Transport

As suggested earlier, carrier-assisted diffusion can sometimes appear to go in an 'uphill' direction, thus giving the impression of active transport. In many animal tissues and micro-organisms, sugars, amino acids, organic acids and vitamins move into the cell up a concentration gradient. This transport is, however, almost completely dependent on having Na^+ or H^+ in the external medium; other ions such as K^+, Rb^+ or Li^+ cannot be substituted. It has been found that the metabolite is carried into the cell along with an ion-carrier complex which is diffusing 'downhill'. In both *Chlorella* and *Neurospora,* glucose is transported in this way along with protons, H^+ (Komor and Tanner, 1974; Slayman and Slayman, 1974).

Co-transport depends on active transport in an indirect way (as indeed does all diffusion) because energy-dependent extrusion pumps ensure that the cytoplasm is kept well below its equilibrium concentration in H^+ and Na^+. These ions tend to diffuse back into the cell and, in doing so, *decrease* their free energy. The energy they give up is coupled, *via* the carrier, to the co-transport of the solute whose free energy is *increased* as it moves into the cell.

Co-transport may also assist the 'uphill' movement of inorganic ions into the cell; it may be a more common process than is generally realized. Recent evidence by Lowendorf *et al.,* (1974) suggests that the active transport of phosphate into the hyphae of *Neurospora* depends on (a) the activity of a proton extrusion pump at the plasmalemma which is sensitive to the pH of the external medium, and (b) the formation of a ternary complex between a proton and a phosphate ion from the external medium with a membrane carrier. The protonated phosphate carrier diffuses to the cytoplasmic side of the membrane *down* the electrochemical gradient of the proton, releasing the proton and phosphate ion into the cytoplasm.

This, and other examples of inorganic ion co-transport suggest a reason why so little progress has been made in elucidating the molecular details of certain 'pumps' particularly of those which transport anions. Put most simply, it may be that these influx pumps do not exist and that the uphill transport is driven by a combination of active extrusion and re-entry by diffusion of protons or perhaps sodium ions.

Co-transport illustrates the ingenious way in which nature can turn necessity to its own advantage. The active excretion of H^+ and Na^+ is essential to maintain pH control and osmo-regulation in the cell, but the energy expended is partly recovered in the transport of essential metabolites and ions into the cell.

MEMBRANE STRUCTURE

It is convenient to discuss membrane structure under two headings; the organization of the *membrane matrix* which is largely a matter of the relationships of the lipid components, and the *substructure* of the protein in the membrane.

THE MEMBRANE MATRIX

Phospholipids

The structure of phospholipid molecules considered earlier provides the key to understanding why it is that the membranes as seen in transverse sections have a characteristic trilaminar appearance. The hydrophilic head regions make hydrogen bonds with water and may become cross-linked to other heads and to proteins through ionic bridges, *e.g.* by calcium ions; thus they are organized into a lattice-like structure. By contrast the long acyl chains of the two fatty acids attached to each phospholipid are strongly hydrophobic, loosely organized and, above their melting point, are relatively fluid. If phospholipid is dispersed in water the 'tails' will take on a conformation which will minimize their contacts with water. The 'heads' will, of course, react favourably with water. If the available water surface is large relative to the amount of lipid, the molecules will arrange themselves as a film-like monolayer with the heads at the water surface and the 'tails' protruding from it at right angles.

If more phospholipid molecules are added to this system so that there are more than can be fitted into a tightly packed monolayer over the water surface, a second type of arrangement occurs quite spontaneously. The phospholipids form two ranks with the heads facing outwards in both and the tails directed inwards to form a non-polar hydrophobic layer sandwiched between them. This bilayer arrangement, which is common to all biological membranes, can also be formed from mixtures of phospholipids under laboratory conditions. The synthetic membranes thus produced have helped in arriving at an understanding of many of the structure/function relationships of natural membranes.

The selected analyses show that a given membrane may contain several types of phospholipid as well as appreciable quantities of sterol. It is probable that there is a great deal more organization of phospholipids in natural membranes than can be demonstrated positively at present. Lipids of one kind may be associated into clumps so that the membrane surface may be very heterogeneous with lipids of differing physical properties arranged in a mosaic.

A mosaic of charged and uncharged areas might occur because some phospholipids carry a net electrostatic charge at normal pH values, *e.g.* phosphatidyl glycerol, while others, like phosphatidyl choline (lecithin) are neutral. This is of significance because it has been shown that, in synthetic bilayers, the surface charge on the phospholipid heads can partly determine both the ion-selectivity and the cation permeability of the membrane (Papahadjopoulos, 1971), and it may also be relevant in determining regions of the surface of the plasmalemma where endo-cytosis may occur.

Local variations in the packing of sterols may render some parts of the membrane less fluid than others and thus determine areas where diffusion may be severely restricted (Papahadjopoulos *et al.,* 1973).

More recently researchers have begun to investigate the possibility that the inner and outer halves of the bilayer may differ in their phospholipid composition. Should this prove to be the case, then it is possible that the barrier properties of the membrane to solute diffusion may be different when the membrane is approached from different sides.

In some special circumstances phospholipid molecules may become arranged into globular micelles in which the polar groups are directed towards the periphery of the sphere, the surface being hydrophilic. This state of affairs can be induced by dehydration in synthetic membrane systems, and in nature by viruses which create membrane instability, *e.g.* sendai virus, and by certain phospholipids (*e.g.* lysolecithin) with wedge-shaped head regions which tend to induce curvature of layers of closely packed phospholipids when they are introduced into a bilayer (Lucy, 1970).

It has been suggested that rapid local transitions from the predominant bilayer to the micellar state are important in membrane fusion and in pinocytosis. If these transitions do occur then it is possible that they may cause transient gaps or pores to be created in the membrane; much physiological evidence points to the conclusion that membranes do have very fine pores in them.

Sterols

The insertion of sterol molecules into the membrane increases the structural order of the hydrophobic region. These molecules lie with their long axes parallel to the hydrocarbon chains of the fatty acids with their more rigid ring structures directed towards the outside and their open chain ends towards the centre.

The mobility of the hydrocarbon chains nearest to the outside of the membrane is, therefore, restricted by these stiffening structures but they remain pliant at their ends so that the central region is fluid (Caspar and

Kirschner, 1971). The rigidity conferred on the membrane by the inclusion of sterols slows down the diffusion of materials through the outer part of the lipid domain in synthetic bilayers (Papahadjopculos *et al.,* 1973).

A Model of the Membrane Matrix

A fact, which it is important to understand but which is difficult to illustrate, is that the centre of the membrane is fluid while the periphery is semi-crystalline. Although it is a stable structure, it is known that phospholipid molecules can be inserted into, and withdrawn from the matrix rapidly and that the structure represents a dynamic steady state when it is part of a biological membrane.

MEMBRANE SUB-STRUCTURE

When a cell or a piece of tissue is frozen and then fractured with a suitable blade, the fracture plane will follow lines of weakness in the structure. Since the central region of the membrane matrix contains no ice it is a potential line of weakness, along which the membrane tends to fracture. Where the fracture line passes tangentially across a cell or organelle, sheets of membrane material become apparent; since membranes tend to be cleaved down the middle it is obvious that the surface exposed is not the true membrane surface but is the membrane interior. A relatively smooth sheet of plasmalemma from an onion root tip on which numerous round particles and depressions can be seen. Some of the particles are arranged in files while others are randomly distributed.

These particles, which are usually 6–9 nm in diameter, are embedded in the membrane and are not resting on its true surface. This was demonstrated by Pinto da Silva and Branton (1970) who etched away the ice from the fracture plane by leaving the specimens under a high vacuum for some minutes after fracturing them. Using this method, the true surface of membranes which lay obliquely to the fracture plane was eventually revealed as the ice from the surrounding cytoplasm sublimed.

The true surface had a much smoother appearance, the undulations of which gave the impression of a blanket lying over the embedded particles. The particles in the membrane can be removed by treatment with proteolytic enzymes and can be re-created in synthetic membranes which have been made in the presence of a hydrophobic protein.

Vail *et al.,* (1974) reported that a synthetic bilayer membrane into which a hydrophobic protein had been incorporated had intercalated particles of 8.5 to 9 nm diameter occupying 12 per cent of the internal membrane surface whereas in bilayers lacking the protein there were none.

The frequency of particles varies according to membrane type; chloroplast and mitochondrial membranes have the highest frequency as one might anticipate from the many metabolic events which are centred upon them. One of the two fracture faces is usually more densely populated than the other. Bearing in mind the amphipathic nature of membrane proteins it is possible that the more densely populated face may reflect the principal orientation of the polar groups of the integral proteins, *i.e.* if most of the polar groups were directed towards the outside of the cell or compartment, the outer fracture face would be the most densely populated. It should be noted that in synthetic lecithin membrane and in nerve myelin, which both have a very low permeability to ions, there are no particles.

A Model of Membrane Structure

The representation of membrane structure is one which commands wide acceptance at the present time. In it we see that large masses of integral protein are inserted to varying extents in the membrane matrix while some are restricted to the surface and others protrude right across the width having a dumbell configuration. The model, which has become known as the *fluid mosaic* (Singer and Nicholson, 1972), has a disorderly appearance in contrast to the neat pictures used to illustrate the unit membrane; the membrane sub-structure has been likened to numerous protein icebergs in a sea of lipid.

Membrane Fluidity

An essential feature of the model is that the various components of the membrane can move relative to one another. Rather than being fixed points, many of the proteins are best thought of as drifting around in the lipid, the viscosity of which will determine the rate at which they move. Evidence that membrane particles can move in the plane of the membrane comes from freeze fracture studies of the plasmalemma of *Mycoplasma mycoides,* in which it was seen that particles became aggregated as the membrane was cooled down but dispersed again when it was warmed up (Rottem *et al.,* 1973). A second line of the evidence comes from studies where two animal cells were induced to fuse with one another (Edidin and Farnbrough, 1973).

One of the cells has a specific antigen in its plasma membrane which the other cell lacked. The distribution of the antigen could be visualized by the binding of a fluorescent antibody.

The fusion of the cells to form a heterocaryon and the subsequent redistribution of the fluorescent antibody were observed using a fluorescence microscope. At first the fluorescent marker was restricted to one half of the heterocaryon, but quite quickly, and in an orderly progression the fluorescence spread right around the coat indicating that the antigenic protein, known to be integrated into the membrane matrix, had diffused in the plane of the membrane. For it to have done this it must have drifted through the lipid. If the heterocaryon was cooled below the transition temperature of its lipids, no mixing of the antigen occurred.

From what we have said above and from earlier comments it is clear that temperature will have a very important influence on the diffusion of materials through, and in the plane of, the membrane because of its effect on the viscosity of the lipid (Edidin, 1974). If the membrane is cooled to the extent that the lipids freeze then such diffusion will become very restricted; the biological significance of this is reflected in the fact that the rates of most physiological processes examined over a range of temperatures are found to have a sharp transition at a temperature which is close to the transition of membrane lipids from a liquid crystalline condition to the gelled condition (Simon, 1974).

Most workers agree that there must be some proteins whose position in the membrane relative to others must be maintained, *e.g.* components of electrontransport chains. Such proteins may require anchoring points either in parts of the membrane which are less fluid than others, or by association with some extra-membrane protein. It is possible that aggregations of sterol molecules might serve to create more viscous patches, and there is evidence, again from synthetic lipid bilayers, that cholesterol can be concentrated in association with certain phospholipids (de Kruyff *et al.,* 1974).

Membrane Synthesis and Flow

The intermediary metabolism concerned with the synthesis of the components of the membrane is beyond the scope of this chapter but it is believed that the components themselves may be centrally assembled and then distributed to the various membranes of the cell. This flow of membrane material can be detected in experiments where cells are provided with a radioactive precursor to a common membrane protein for a short while, and then returned to non-radio-active medium. This type of analysis has not been performed on plant cells but in animal cells, harvested at intervals after pulse-labelling, the radioactive label appears first in the endoplasmic reticulum and then in the Golgi cisternae. Subsequently, the radioactivity of these compartments declines followed by an increase in label associated with the plasmamembrane some hours later. Since mitochondria and chloroplasts probably do not have all of the metabolic apparatus to assemble their own membranes it is thought that they too may obtain partly finished membranes from the endoplasmic reticulum.

There is evidence that the cytoplasm contains numerous membrane-bounded vesicles which frequently appear to be in conjunction with the major membranes of the cell (*e.g.* Mahlberg *et al.,* 1974). They are particularly prominent in cells synthesizing walls and, since it is known that precursors of wall synthesis are formed inside the cell and elaborated outside, it is reasonable to conclude that the vesicles contain precursors which are discharged after fusion with the plasmalemma (Heyn, 1971). It has also been found that particulate material from the external medium can be detected in vesicles within the cytoplasm (Mayo and Cocking, 1969; Robards and Robb, 1974). These results suggest that vesicles can both fuse with, and be formed from the plasmalemma and other membranes, thus allowing materials to pass out of or into the cell without their having to cross a membrane; such movements are known as exoand endo-cytosis, respectively.

Since adjacent membranes can frequently be found to be in contact over comparatively large areas and yet show no tendency to fuse with one another, it is believed that special proteins or phospholipids (*e.g.* lysolecithin) in the vesicle membrane may trigger fusion where they make contact with the larger membrane sheet.

CORRELATION OF STRUCTURE AND FUNCTION

Having a picture in mind of the way membranes are assembled and of the driving forces which operate across them it is now possible to consider how their design, particularly that of the plasmalemma, manages to reconcile two conflicting sets of priorities.

This conflict may be summarized as follows: the cell usually contains solutes at a far higher concentration than in the external medium; without any barrier these solutes would disperse by diffusion.

Many solutes are accumulated by mechanisms which consume metabolic energy and it is sensible for nature to minimize their subsequent leakage and thereby economize in the use of 'fuel'. Insulation of this type might be most effective if the cell were surrounded by a barrier impenetrable to virtually everything except water and dissolved gases. Such a barrier would, however, isolate the cell from the outside world and render it insensitive to stimulae and incapable of growth, it might eliminate its interactions with other cells in adjacent or remote tissues and prevent the excretion of harmful excesses of substances generated during metabolism.

The cell membrane must provide a balance, therefore, between minimizing losses and permitting a selective interchange of substances between the cell interior and the surroundings. The balance is achieved, in many cases, by providing channels or other more elaborate carrier mechanisms whereby materials crossing the membrane may essentially by-pass the hydrophobic region.

MEMBRANE PORES AND CHANNELS

Lipid or oil is a very effective electrical insulator (it is used in high voltage underground electric cables for this purpose) and it is not surprising to find that the electrical resistance of synthetic bilayers made from pure phospholipid is very high, being 10^7 to 10^9 ohms cm^2. In nature, the current conducted across membranes is carried by ions and we can see that an unmodified lipid bilayer with such a high resistance is a poor material across which to conduct this essential current of electrolyte. It is not surprising to find, therefore that the electrical resistance of natural membranes is very much less than that of bilayers, usually lying in a wide range of 10^2 to 10^5 ohms cm^2.

As described lipophilic carriers can greatly reduce the resistance (and hence increase the c onductance) of synthetic membranes and may resemble the carrier molecules in membranes, but it is also widely believed that membranes contain water-filled pores which must contribute to their relatively high conductance by allowing water and selected solutes to by-pass the lipid domain of the membrane.

Certain antibiotic molecules, such as nystatin, appear to condense cholesterol molecules in both synthetic (Holz and Finkelstein, 1970) and natural membranes (de Kruijff and Demel, 1974) to form pores with a radius of *ca.* 0.4 nm. The presence of such pores greatly increases the electrical conductance and hydraulic conductivity of the membrane.

In passing we might note that many antibiotics have their effect by enormously increasing the passive permeability of cell membranes causing non-resistant cells to lose their contents or to lyse. Interest in these substances stems from the experimental evidence that cell membranes also possess pores of similar size and that the antibiotic merely induces an extreme expresssion of the normal condition.

Although the word 'pore' is often used to describe channels through which solutes and water can move, we should resist the temptation to conclude that all 'pores' are definable structural entities like the ones induced by nystatin. In many instances a 'pore' may be more like a transient imperfection in membrane structure. This latter type may have a certain statistical probability but have no fixed position.

Proteins which traverse the lipid layer may give rise to hydrophilic channels or pores. Indirect evidence supporting this idea comes from a study in which red blood cell membranes were exposed by deeply etching frozen cells under vacuum (Pinto da Silva, 1973). Shrinkage of the membrane surface was observed in areas overlying groups of membrane particles possibly due to sublimation of ice from within the embedded particles; for this to have occurred this water would be a free liquid in the thawed condition.

The first circumstantial evidence for the existence of pores come from studies by Collander and Barlund (1933) on the permeation of the giant internodal cells of the alga, *Chara ceratophylla,* by a number of uncharged solutes and water. They found that in almost every case the rate of movement of a substance into a cell depended on its molecular weight and dimensions and on its solubility in oil relative to water; substances with high oil solubility permeated most rapidly.

This strongly suggested that movement across cell membranes involved the movement of the solute out of the water, its solution in lipid and subsequent diffusion through it, and its re-entry into the aqueous phase at the inside face of the membrane. The authors found, however, that several small molecules permeated the membrane far more rapidly (more than 100 times faster in the case of water) than their relative solubility in oil suggested.

The upper size limit for molecules which behaved anomalously was a radius of 0.4 nm and it was suggested that the membrane was constructed as a very fine sieve containing pores of 0.4 nm radius through which water (0.25 nm radius) and certain solutes could move.

Since this early work a great deal has been done and the equivalent pore radius in many plasma membranes has been confirmed as 0.4 nm. It must be said, however, that some authorities are reluctant to accept that pores can provide channels for the bulk movement of water and solutes; the arguments for and against pores have been clearly discussed in Oschman *et al.,* (1974).

There has been much discussion about whether pores admit ions, and if so, which ones. Membranes are known to control very precisely the relative rates at which various ions will diffuse across them; potassium ions will diffuse ten to one-hundred times faster than sodium. Ions in solution are hydrated by binding one or more water molecules; it requires a great deal of energy to dehydrate an ion and for this reason we should think of ions in all natural circumstances as being hydrated.

Sodium binds 5 water molecules, whereas potassium binds 3, the former is, therefore, the more bulky ion whose diffusion into the narrow water filled pores would be slower than for the smaller hydrated potassium ion.

On the other hand the anion chloride, which has only one water molecule in its hydration shell, diffuses much more slowly than either K^+ or Na^+.

This is probably due to the fact that 'pores' carry a predominantly negative charge so that cations would be attracted to them, while anions would be repelled—a small number of positively charged pores would handle the flow of anions. Polyvalent cations and anions have much more water in their hydration shells, *e.g.* Ca^{2+} has 10 and $SO_4{}^{2-}$ has 8 and it seems likely that these would be totally excluded from the pores.

ION PUMPS AND MEMBRANE SUBSTRUCTURE

It is most unlikely that the intercalated membrane particles described earlier are simple ion-carriers—they are much too large—but they may be ion pumps, or groups of pumps. The best evidence for this assertion comes from work on the ATPase which pumps Ca^{2+} across the membranes of sarcoplasmic reticulum.

Racker (1972) was able to isolate this enzyme and put it back into a synthetic bilayer membrane made from soybean phospholipids, and thus reconstitute a membrane which could actively transport Ca^{2+} when ATP and Mg^{2+} were present in the medium. Working with a slightly different system Packer *et al.,* (1974) showed that freeze etched fractures of both the original tissue membrane and reconstituted membrane containing the purified ATPase were densely studded by numerous particles 8.5 nm in diameter. There can be little doubt that the particles were the calcium-pumping ATPase. Similar, less complete, evidence is also available for the Na^+/K^+ -dependent ATPase of the red blood cell. There is no insuperable difficulty in repeating such observations with plant tissues but, at the time of writing there is no report that this has been done.

One should not conclude that all of the membrane substructure is concerned with solute transport, especially in mitochondria and chloroplasts where other biochemical activities are centred on membranes.

ENDOCYTOSIS AND VESICULAR TRANSPORT

The presence of abundant vesicles in the vicinity of the plasmalemma of plant cells has been pointed out earlier (p. 44); however, it is not yet clear what contribution endocytosis makes to the total solute transport across this, or any other membrane.

There is kinetic evidence from several sources which is consistent with the notion that a measurable fraction of various ions in the cytoplasm of plant cells is sequestered into a compartment separate from the bulk of the cytoplasm. MacRobbie (1969) found that labelled Cl^- transported into the vacuole of *Nitella translucens* could be easily resolved into a fast and a slow component. The fast component was envisaged as being delivered to the vacuole formed in vesicles which had been found at the plasmalemma.

The chloride ions in these vesicles would not have mixed with the unlabelled chloride in the bulk cytoplasm, whereas labelled Cl^- delivered directly to the cytoplasm *via a* pumping mechanism would have to mix with a much larger pool of unlabelled chloride ions. At the beginning of the experiment, therefore, the labelled Cl^- in the vacuole increased more rapidly than expected. There are other examples of this kind discussed in MacRobbie (1971) and Baker and Hall (1973).

If the vesicles contained nothing more than a small volume of the outside solution, their net contribution to the solute content of the cell would be very small and they could not exercise the ion-selectivity which characterizes cell membranes.

Baker and Hall (1973) suggest that endocytosis becomes a much more plausible transport process if it is assumed that ions become selectively bound to the membrane surface and thus become concentrated from the dilute external medium. This assumption is entirely reasonable because both proteins and phospholipid heads can provide ion-selective binding sites.

These authors also relate membrane-bound ATPase activity with the subsequent invagination of the plasmalemma to form a vesicle but this implies that vesicle formation is energy-dependent. This is against the substantial evidence from animal cells which shows that micro-vesicle formation results from Brownian movement of the membrane and does not need to be energized by ATP (Casely-Smith, 1969); ATP is required for the very large vesicles formed by re-orientation of micro- fibrils in phagocytosis, but the type of vesicle under discussion here is much smaller, probably 70–100 nm in diameter and is not dependent on microfibrils. There is visual evidence that these micro-vesicles can discharge their contents directly into lysosomal vacuoles (Casely-Smith and Chin, 1971).

The important advance of thinking required to deal with endocytosis will provide a severe strain on biophysical theorists accustomed to thinking of cells as homogeneous phases separated by diffusion barriers, partly because materials can reach the interior of the cell without crossing a membrane at all.

Although endocytosis is probably a common phenomenon it is likely to remain an 'unknown quantity' until new experimental techniques are devised.

8

Photosynthesis

Photosynthesis is a process that converts carbon dioxide into organic compounds, especially sugars, using the energy from sunlight. Photosynthesis occurs in plants, algae, and many species of bacteria, but not in archaea. Photosynthetic organisms are called photoautotrophs, since they can create their own food. In plants, algae, and cyanobacteria, photosynthesis uses carbon dioxide and water, releasing oxygen as a waste product. Photosynthesis is vital for all aerobic life on Earth.

As well as maintaining the normal level of oxygen in the atmosphere, nearly all life either depends on it directly as a source of energy, or indirectly as the ultimate source of the energy in their food (the exceptions are chemoautotrophs that live in rocks or around deep sea hydrothermal vents). The rate of energy capture by photosynthesis is immense, approximately 100 terawatts, which is about six times larger than the power consumption of human civilization. As well as energy, photosynthesis is also the source of the carbon in all the organic compounds within organisms' bodies. In all, photosynthetic organisms convert around 100–115 teragrams of carbon into biomass per year.

Although photosynthesis can happen in different ways in different species, some features are always the same. For example, the process always begins when energy from light is absorbed by proteins called photosynthetic reaction centers that contain chlorophylls. In plants, these proteins are held inside organelles called chloroplasts, while in bacteria they are embedded in the plasma membrane. Some of the light energy gathered by chlorophylls is stored in the form of adenosine triphosphate (ATP). The rest of the energy is used to remove electrons from a substance such as water. These electrons are then used in the reactions that turn carbon dioxide into organic compounds. In plants, algae and cyanobacteria, this is done by a sequence of reactions called the Calvin cycle, but different sets of reactions are found in some bacteria, such as the reverse Krebs cycle in Chlorobium. Many photosynthetic organisms have adaptations that concentrate or store carbon dioxide. This helps reduce a wasteful process called photorespiration that can consume part of the sugar produced during photo-synthesis. The first photosynthetic organisms probably evolved about 3,500 million years ago, early in the evolutionary history of life, when all forms of life on Earth were microorganisms and the atmosphere had much more carbon dioxide. They most likely used hydrogen or hydrogen sulfide as sources of electrons, rather than water. Cyanobacteria appeared later, around 3,000 million years ago, and drastically changed the Earth when they began to oxygenate the atmosphere, beginning about 2,400 million years ago.

This new atmosphere allowed the evolution of complex life such as protists. Eventually, no later than a billion years ago, one of these protists formed a symbiotic relationship with a cyanobacterium, producing the ancestor of many plants and algae. The chloroplasts in modern plants are the descendants of these ancient symbiotic cyanobacteria.

Photosynthetic organisms are photoautotrophs, which means that they are repositories of energy, they are able to synthesize food directly from carbon dioxide using energy from light.

They accrue it as part of their potential energy. However, not all organisms that use light as a source of energy carry out photosynthesis, since photoheterotrophs use organic compounds, rather than carbon dioxide, as a source of carbon. In plants, algae and cyanobacteria, photosynthesis releases oxygen.

This is called oxygenic photosynthesis. Although there are some differences between oxygenic photosynthesis in plants, algae and cyanobacteria, the overall process is quite similar in these organisms. However, there are some types of bacteria that carry out anoxygenic photosynthesis, which consumes carbon dioxide but does not release oxygen.

Carbon dioxide is converted into sugars in a process called carbon fixation. Carbon fixation is a redox reaction, so photosynthesis needs to supply both a source of energy to drive this process, and the electrons needed to convert carbon dioxide into a carbohydrate, which is a reduction reaction. In general outline, photosynthesis is the opposite of cellular respiration, where glucose and other compounds are oxidized to produce carbon dioxide, water, and release chemical energy. However, the two processes take place through a different sequence of chemical reactions and in different cellular compartments.

The general equation for photosynthesis is therefore:

$$2n\ CO_2 + 2n\ H_2O + \text{photons} \rightarrow 2(CH_2O)_n + n\ O_2 + 2n\ A$$

Carbon dioxide + electron donor + light energy → carbohydrate + oxygen + oxidized electron donor

Since water is used as the electron donor in oxygenic photosynthesis, the equation for this process is:

$$2n\ CO_2 + 2n\ H_2O + \text{photons} \rightarrow 2(CH_2O)_n + 2n\ O_2$$

carbon dioxide + water + light energy → carbohydrate + oxygen

Other processes substitute other compounds (such as arsenite) for water in the electron–supply role; the microbes use sunlight to oxidize arsenite to arsenate: The equation for this reaction is:

$$(AsO_3^{3-}) + CO_2 + \text{photons} \rightarrow CO + (AsO_4^{3-})$$

arsenite + carbon dioxide + light energy → carbon monoxide + arsenate.

Photosynthesis occurs in two stages. In the first stage, light–dependent reactions or light reactions capture the energy of light and use it to make the energy–storage molecules ATP and NADPH. During the second stage, the light–independent reactions use these products to capture and reduce carbon dioxide. Most organisms that utilize photosynthesis to produce oxygen use visible light to do so, although at least three use infrared radiation.

LIGHT, LEAVES, AND PHOTOSYNTHESIS

Scaling up from the chloroplast to the leaf adds new levels of complexity to photosynthesis. At the same time, the structural and functional properties of the leaf make possible other levels of regulation. We will start by examining how leaf anatomy, and movements by chloroplasts and leaves, control the absorption of light for photosynthesis.

Then we will describe how chloroplasts and leaves adapt to their light environment and how the photosynthetic response of leaves grown under low light reflects their adaptation to a low-light environment. Leaves also adapt to high light conditions, illustrating that plants are physiologically flexible and that they adapt to their immediate environment. Both the amount of light and the amount of CO_2 determine the photosynthetic response of leaves. In some situations, photosynthesis is limited by an inadequate supply of light or CO_2.

In other situations, absorption of too much light can cause severe problems, and special mechanisms protect the photosynthetic system from excessive light. Multiple levels of control over photosynthesis allow plants to grow successfully in a constantly changing environment and different habitats.

CONCEPTS AND UNITS IN THEMEASUREMENT OF LIGHT

Three light parameters are especially important in the measurement of light:

- Spectral quality,
- Amount, and
- Direction.

Adiscussion of the amount and direction of light reaching the plant requires consideration of the geometry of the part of the plant that receives the light: Is the plant organ flat or cylindrical? Flat, or planar, light sensors are best suited for flat leaves.

The light reaching the plant can be measured as energy, and the amount of energy that falls on a flat sensor of known area per unit time is quantified as irradiance. Units can be expressed in terms of energy, such as watts per square meter (Wm^{-2}). Time (seconds) is contained within the term watt: 1 W = 1 joule (J) s^{-1}. Light can also be measured as the number of incident quanta (singular *quantum*). In this case, units can be expressed in moles per square meter per second (mol m^{-2} s^{-1}), where *moles* refers to the number of photons (1 mol of light = 6.02×10^{23} photons, Avogadro's number). This measure is called photon irradiance. Quanta and energy units can be interconverted relatively easily, provided that the wavelength of the light, λ, is known.

The energy of a photon is related to its wavelength as follows:

$$E = \frac{hc}{\lambda}$$

where c is the speed of light (3×10^8 ms^{-1}), h is Planck's constant (6.63×10^{-34} Js), and λ is the wavelength of light, usually expressed in nm (1 nm = 10^{-9} m). From this equation it can be shown that a photon at 400 nm has twice the energy of a photon at 800 nm. Now let's turn our attention to the direction of light. Light can strike a flat surface directly from above or obliquely. When light deviates from perpendicular, irradiance is proportional to the cosine of the angle at which the light rays hit the sensor.

There are many examples in nature in which the lightintercepting object is not flat (*e.g.*, complex shoots, whole plants, chloroplasts). In addition, in some situations light can come from many directions simultaneously (*e.g.*, direct light from the sun plus the light that is reflected upward from sand, soil, or snow). In these situations it makes more sense to measure light with a spherical sensor that takes measurements omnidirectionally (from all directions). The term for this omnidirectional measurement is fluence rate, and this quantity can be expressed in watts per square meter (W m^{-2}) or moles per square meter per second (mol m^{-2} s^{-1}). The units clearly indicate whether light is being measured as energy (W) or as photons (mol).

In contrast to a flat sensor's sensitivity, the sensitivity to light of a spherical sensor is independent of direction. Depending on whether the light is collimated (rays are parallel) or diffuse (rays travel in random directions), values for fluence rate versus irradiance measured with a flat or a spherical sensor can provide different values.

Photosynthetically Active Radiation

PAR, 400–700 nm may also be expressed in terms of energy (W m^{-2}) or quanta (mol m^{-2} s^{-1}). Note that PAR is an irradiance-type measurement. In research on photosynthesis, when PAR is expressed on a quantum basis, it is given the special term photosynthetic photon flux density (PPFD). However, it has been suggested that the term *density* be discontinued because within the International System of Units (SI units, where *SI* stands for *System International*), *density* can mean area or volume. In summary, when choosing how to quantify light, it is important to match sensor geometry and spectral response with that of the plant. Flat, cosine-corrected sensors are ideally suited to measure the amount of light that strikes the surface of a leaf; spherical sensors are more appropriate in other situations, such as in studies of a chloroplast suspension or a branch from a tree.

How much light is there on a sunny day, and what is the relationship between PAR irradiance and PAR fluence rate? Under direct sunlight, PAR irradiance and fluence rate are both about 2000 μmol m^{-2} s^{-1}, though higher values can be measured at high altitudes. The corresponding value in energy units is about 400 W m^{-2}.

Leaf Anatomy Maximizes Light Absorption

Roughly 1.3 kW m^{-2} of radiant energy from the sun reaches Earth, but only about 5 per cent of this energy can be converted into carbohydrates by a photosynthesizing leaf. The reason this percentage is so low is that a major fraction of the incident light is of a wavelength either too short or too long to be absorbed by the photosynthetic pigments. Of the absorbed light energy, a significant fraction is lost as heat, and a smaller amount is lost as fluorescence. Radiant energy from the sun consists of many different wavelengths of light. Only photons of wavelengths from 400 to 700 nm are utilized in photosynthesis, and about 85 to 90 per cent of this PAR is absorbed by the leaf; the remainder is either reflected at the leaf surface or transmitted through the leaf. Because chlorophyll absorbs very strongly in the blue and the red regions of the spectrum, the transmitted and reflected light are vastly enriched in green—hence the green colour of vegetation.

The anatomy of the leaf is highly specialized for light absorption. The outermost cell layer, the epidermis, is typically transparent to visible light, and the individual cells are often convex. Convex epidermal cells can act as lenses and can focus light so that the amount reaching some of the chloroplasts can be many times greater than the amount of ambient light. Epidermal focusing is common among herbaceous plants and is especially prominent among tropical plants that grow in the forest understory, where light levels are very low.

Below the epidermis, the top layers of photosynthetic cells are called palisade cells; they are shaped like pillars that stand in parallel columns one to three layers deep. Some leaves have several layers of columnar palisade cells, and we may wonder how efficient it is for a plant to invest energy in the development of multiple cell layers when the high chlorophyll content of the first layer would appear to allow little transmission of the incident light to the leaf interior. In fact, more light than might be expected penetrates the first layer of palisade cells because of the sieve effect and light channeling.

The sieve effect is due to the fact that chlorophyll is not uniformly distributed throughout cells but instead is confined to the chloroplasts. This packaging of chlorophyll results in shading between the chlorophyll molecules and creates gaps between the chloroplasts, where light is not absorbed—hence the reference to a sieve. Because of the sieve effect, the total absorption of light by a given amount of chlorophyll in a palisade cell is less than the light absorbed by the same amount of chlorophyll in a solution.

Light channeling occurs when some of the incident light is propagated through the central vacuole of the palisade cells and through the air spaces between the cells, an arrangement that facilitates the transmission of light into the leaf interior.

Below the palisade layers is the spongy mesophyll, where the cells are very irregular in shape and are surrounded by large air spaces. The large air spaces generate many interfaces between air and water that reflect and refract the light, thereby randomizing its direction of travel. This phenomenon is called *light scattering*. Light scattering is especially important in leaves because the multiple reflections between cell–air interfaces greatly increase the length of the path over which photons travel, thereby increasing the probability for absorption. In fact, photon path lengths within leaves are commonly four times or more longer than the thickness of the leaf. Thus the palisade cell properties that allow light to pass through, and the spongy mesophyll cell properties that are conducive to light scattering, result in more uniform light absorption throughout the leaf.

Some environments, such as deserts, have so much light that it is potentially harmful to leaves. In these environments leaves often have special anatomic features, such as hairs, salt glands, and epicuticular wax that increase the reflection of light from the leaf surface, thereby reducing light absorption. Such adaptations can decrease light absorption by as much as 40 per cent, minimizing heating and other problems associated with the absorption of too much light.

Chloroplast Movement and LeafMovement can Control Light Absorption

Chloroplast movement is widespread among algae, mosses, and leaves of higher plants. If chloroplast orientation and location are controlled, leaves can regulate how much of the incident light is absorbed. Under low light, chloroplasts gather at the cell surfaces parallel to the plane of the leaf so that they are aligned perpendicularly to the incident light—a position that maximizes absorption of light.

Under high light, the chloroplasts move to the cell surfaces that are parallel to the incident light, thus avoiding excess absorption of light. Such chloroplast rearrangement can decrease the amount of light absorbed by the leaf by about 15 per cent. Chloroplast movement in leaves is a typical blue-light response. Blue light also controls chloroplast orientation in many of the lower plants, but in some algae, chloroplast movement is controlled by phytochrome. In leaves, chloroplasts move along actin microfilaments in the cytoplasm, and calcium regulates their movement. Leaves have the highest light absorption when the leaf blade, or lamina, is perpendicular to the incident light. Some plants control light absorption by solar tracking; that is, their leaves continuously adjust the orientation of their laminae such that they remain perpendicular to the sun's rays. Alfalfa, cotton, soybean, bean, lupine, and some wild species of the mallow family (Malvaceae) are examples of the numerous plant species that are capable of solar tracking.

Solar-tracking leaves keep a nearly vertical position at sunrise, facing the eastern horizon, where the sun will rise. The leaf blades then lock on to the rising sun and follow its movement across the sky with an accuracy of ±15° until sunset, when the laminae are nearly vertical, facing the west, where the sun will set. During the night the leaf takes a horizontal position and reorients just before dawn so that it faces the eastern horizon in anticipation of another sunrise. Leaves track the sun only on clear days, and they stop when a cloud obscures the sun.

In the case of intermittent cloud cover, some leaves can reorient as rapidly as 90° per hour and thus can catch up to the new solar position when the sun emerges from behind a cloud. Solar tracking is another blue-light response, and the sensing of blue light in solar-tracking leaves occurs in specialized regions. In species of *Lavatera* (Malvaceae), the photosensitive region is located in or near the major leaf veins. In lupines, (*Lupinus*, Fabaceae), leaves consist of five or more leaflets, and the photosensitive region is located in the basal part of each leaflet lamina. In many cases, leaf orientation is controlled by a specialized organ called the pulvinus (plural *pulvini*), found at the junction between the blade and petiole.

The pulvinus contains motor cells that change their osmotic potential and generate mechanical forces that determine laminar orientation. In other plants, leaf orientation is controlled by small mechanical changes along the length of the petiole and by movements of the younger parts of the stem. Some solar-tracking plants can also move their leaves such that they avoid full exposure to sunlight, thus minimizing heating and water loss. Building on the term heliotropism (bending towards the sun), which is often used to describe sun-induced leaf movements, these sunavoiding leaves are called *paraheliotropic*, and leaves that maximize light interception by solar tracking are called *diaheliotropic*.

Some plant species can display diaheliotropic movements when they are well watered and paraheliotropic movements when they experience water stress. Since full sunlight usually exceeds the amount of light that can be utilized for photosynthesis, what advantage is gained by solar tracking? By keeping leaves perpendicular to the sun, solar-tracking plants maintain maximum photosynthetic rates throughout the day, including early morning and late afternoon.

Moreover, air temperature is lower during the early morning and late afternoon, so water stress is lower. Solar tracking therefore gives an advantage to plants that grow in arid regions.

Plants Adapt to Sun and Shade

Some plants have enough developmental plasticity to adapt to a range of light regimes, growing as sun plants in sunny areas and as shade plants in shady habitats. Some shady habitats receive less than 1 per cent of the PAR available in an exposed habitat. Leaves that are adapted to very sunny or very shady environments are often unable to survive in the other type of habitat.

Sun and shade leaves have some contrasting characteristics:

- *Shade leaves* have more total chlorophyll per reaction centre, have a higher ratio of chlorophyll *b* to chlorophyll *a*, and are usually thinner than sun leaves.
- *Sun leaves* have more rubisco, and a larger pool of xanthophyll cycle components than shade leaves.

Contrasting anatomic characteristics can also be found in leaves of the same plant that are exposed to different light regimes. Some anatomic differences between a leaf grown in the sun and a leaf grown in the shade. Sun-grown leaves are thicker and have longer palisade cells than leaves growing in the shade. Even different parts of a single leaf show adaptations to their light microenvironment. Cells in the upper surface of the leaf, which are exposed to the highest prevailing photon flux, have characteristics of cells from leaves grown in full sunlight; cells in the lower surface of the leaf have characteristics of cells found in shade-grown leaves.

These morphological and biochemical modifications are associated with specific functions. Far-red light is absorbed primarily by PSI, and altering the ratio of PSI to PSII or changing the light-harvesting antennae associated with the photosystems makes it possible to maintain a better balance of energy flow through the two photosystems. These adaptations are found in nature; some shade plants show a 3:1 ratio of photosystem II to photosystem I reaction centers, compared with the 2:1 ratio found in sun plants. Other shade plants, rather than altering the ratio of PSI to PSII, add more antennae chlorophyll to PSII.

These adaptations appear to enhance light absorption and energy transfer in shady environments, where far-red light is more abundant. Sun and shade plants also differ in their respiration rates, and these differences alter the relationship between respiration and photosynthesis.

Plants Compete for Sunlight

Plants normally compete for sunlight. Held upright by stems and trunks, leaves configure a canopy that absorbs light and influences photosynthetic rates and growth beneath them. Leaves that are shaded by other leaves have much lower photosynthetic rates.

Some plants have very thick leaves that transmit little, if any, light. Other plants, such as those of the dandelion (*Taraxacum* sp.), have a rosette growth habit, in which leaves grow radially very close to each other and to the stem, thus preventing the growth of any leaves below them. Trees represent an outstanding adaptation for light interception. The elaborate branching structure of trees vastly increases the interception of sunlight. Very little PAR penetrates the canopy of many forests; almost all of it is absorbed by leaves. Another feature of the shady habitat is sunflecks, patches of sunlight that pass through small gaps in the leaf canopy and move across shaded leaves as the sun moves. In a dense forest, sunflecks can change the photon flux impinging on a leaf in the forest floor more than tenfold within seconds.

For some of these leaves, a sunfleck contains nearly 50 per cent of the total light energy available during the day, but this critical energy is available for only a few minutes in a very high dose. Sunflecks also play a role in the carbon metabolism of lower leaves in dense crops that are shaded by the upper leaves of the plant. Rapid responses of both the photosynthetic apparatus and the stomata to sunflecks have been of substantial interest to plant physiologists and ecologists because they represent unique physiological responses specialized for capturing a short burst of sunlight.

LEAVES AND LEAF STRUCTURE

Plants are the only photosynthetic organisms to have leaves (and not all plants have leaves). A leaf may be viewed as a solar collector crammed full of photosynthetic cells. The raw materials of photosynthesis, water and carbon dioxide, enter the cells of the leaf, and the products of photosynthesis, sugar and oxygen, leave the leaf.

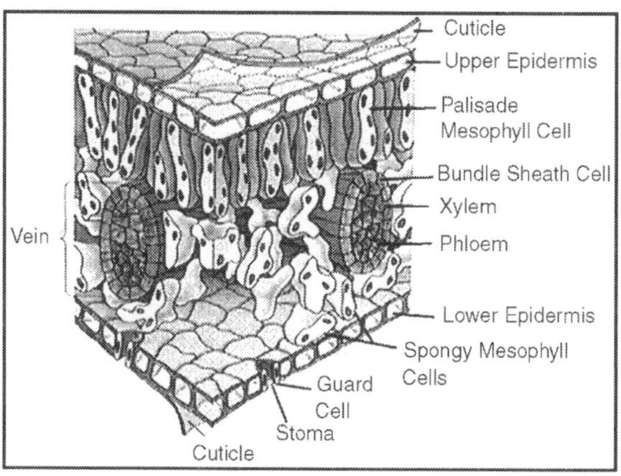

Fig. Cross Section of a Leaf, Showing the Anatomical Features Important to the Study of Photosynthesis: stoma, Guard Cell, Mesophyll Cells, and Vein.

Water enters the root and is transported up to the leaves through specialized plant cells known as xylem (pronounces zigh–lem). Land plants must guard against drying out (desiccation) and so have evolved specialized structures known as stomata to allow gas to enter and leave the leaf. Carbon dioxide cannot pass through the protective waxy layer covering the leaf (cuticle), but it can enter the leaf through an opening (the stoma; plural = stomata; Greek for hole) flanked by two guard cells. Likewise, oxygen produced during photosynthesis can only pass out of the leaf through the opened stomata. Unfortunately for the plant, while these gases are moving between the inside and outside of the leaf, a great deal water is also lost. Cottonwood trees, for example, will lose 100 gallons of water per hour during hot desert days. Carbon dioxide enters single–celled and aquatic autotrophs through no specialized structures.

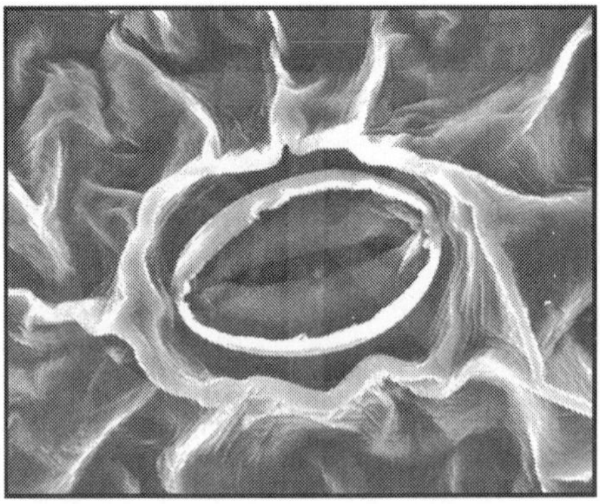

Fig. Pea Leaf Stoma

The Nature of Light

White light is separated into the different colours (=wavelengths) of light by passing it through a prism. Wavelength is defined as the distance from peak to peak (or trough to trough). The energy of is inversely porportional to the wavelength: longer wavelengths have less energy than do shorter ones.

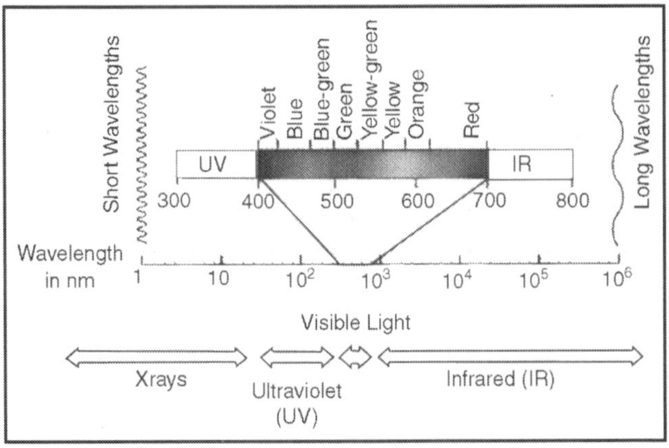

The order of colours is determined by the wavelength of light. Visible light is one small part of the electromagnetic spectrum. The longer the wavelength of visible light, the more red the colour. Likewise the shorter wavelengths are towards the violet side of the spectrum. Wavelengths longer than red are referred to as infrared, while those shorter than violet are ultraviolet.

Light behaves both as a wave and a particle. Wave properties of light include the bending of the wave path when passing from one material (medium) into another (*i.e.* the prism, rainbows, pencil in a glass–of–water, etc.).

The particle properties are demonstrated by the photoelectric effect. Zinc exposed to ultraviolet light becomes positively charged because light energy forces electrons from the zinc. These electrons can create an electrical current. Sodium, potassium and selenium have critical wavelengths in the visible light range. The critical wavelength is the maximum wavelength of light (visible or invisible) that creates a photoelectric effect.

Chlorophyll and Accessory Pigments

A pigment is any substance that absorbs light. The colour of the pigment comes from the wavelengths of light reflected (in other words, those not absorbed). Chlorophyll, the green pigment common to all photosynthetic cells, absorbs all wavelengths of visible light except green, which it reflects to be detected by our eyes.

Black pigments absorb all of the wavelengths that strike them. White pigments/lighter colours reflect all or almost all of the energy striking them. Pigments have their own characteristic absorption spectra, the absorption pattern of a given pigment.

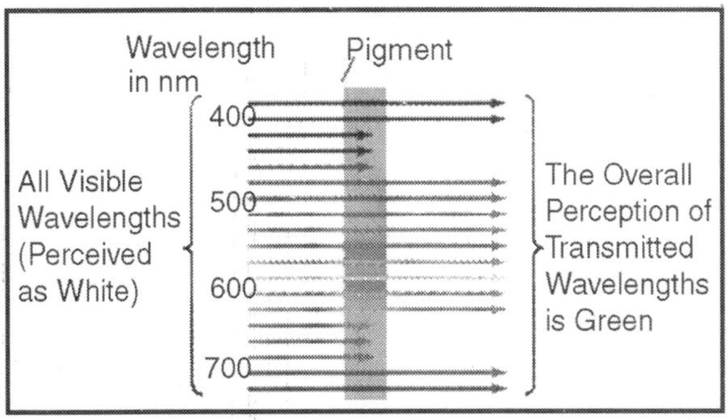

Fig. Absorption and Transmission of Different Wavelengths of Light by a Hypothetical Pigment.

Chlorophyll is a complex molecule. Several modifications of chlorophyll occur among plants and other photosynthetic organisms. All photosynthetic organisms (plants, certain protistans, prochlorobacteria, and cyanobacteria) have chlorophyll a. Accessory pigments absorb energy that chlorophyll a does not absorb.

Accessory pigments include chlorophyll b (also c, d, and e in algae and protistans), xanthophylls, and carotenoids (such as beta–carotene). Chlorophyll a absorbs its energy from the Violet–Blue and Reddish orange–Red wavelengths, and little from the intermediate (Green–Yellow–Orange) wavelengths.

Carotenoids and chlorophyll b absorb some of the energy in the green wavelength. Why not so much in the orange and yellow wavelengths? Both chlorophylls also absorb in the orange–red end of the spectrum (with longer wavelengths and lower energy).

The origins of photosynthetic organisms in the sea may account for this. Shorter wavelengths (with more energy) do not penetrate much below 5 meters deep in sea water. The ability to absorb some energy from the longer (hence more penetrating) wavelengths might have been an advantage to early photosynthetic algae that were not able to be in the upper (photic) zone of the sea all the time.

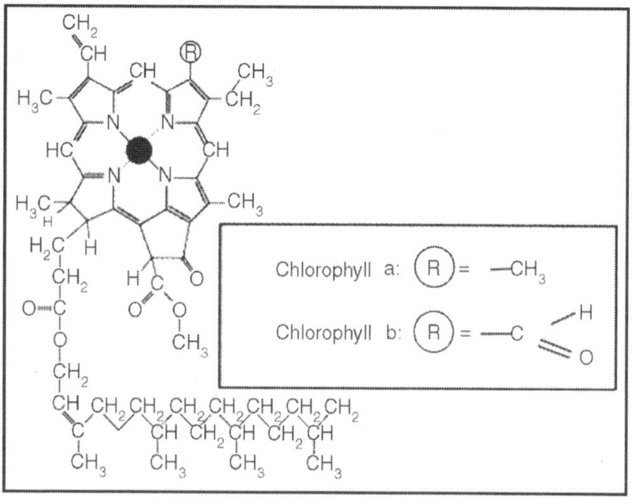

Fig. The Molecular Structure of Chlorophylls.

The action spectrum of photosynthesis is the relative effectiveness of different wavelengths of light at generating electrons. If a pigment absorbs light energy, one of three things will occur. Energy is dissipated as heat. The energy may be emitted immediately as a longer wavelength, a phenomenon known as fluorescence. Energy may trigger a chemical reaction, as in photosynthesis.

Chlorophyll only triggers a chemical reaction when it is associated with proteins embedded in a membrane (as in a chloroplast) or the membrane infoldings found in photosynthetic prokaryotes such as cyanobacteria and prochlorobacteria.

Structure of the Chloroplastand Photosynthetic Membranes

The thylakoid is the structural unit of photosynthesis. Both photosynthetic prokaryotes and eukaryotes have these flattened sacs/vesicles containing photosynthetic chemicals. Only eukaryotes have chloroplasts with a surrounding membrane. Thylakoids are stacked like pancakes in stacks known collectively as grana. The areas between grana are referred to as stroma. While the mitochondrion has two membrane systems, the chloroplast has three, forming three compartments.

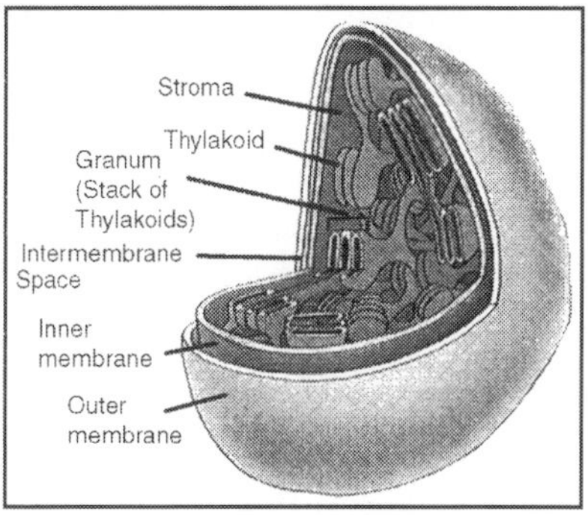

Fig. Structure of a Chloroplast.

PHOTOSYNTHETIC RESPONSES TO LIGHT BY THE INTACT LEAF

Light is a critical resource for plants that can often limit growth and reproduction. The photosynthetic properties of the leaf provide valuable information about plant adaptations to their light environment. We describe typical photosynthetic responses to light as measured in light-response curves. We also consider how an important feature of light-response curves, the light compensation point, explains contrasting physiological properties of sun and shade plants. We then describe quantum yields of photosynthesis in the intact leaf, and the differences in quantum yields between C_3 and C_4 plants. The section closes with descriptions of leaf adaptations to excess light, and the different pathways of heat dissipation in the leaf.

Light-Response Curves RevealPhotosynthetic Properties

Measuring CO_2 fixation in intact leaves at increasing photon flux allows us to construct light-response curves that provide useful information about the photosynthetic properties of leaves. In the dark there is no photosynthetic carbon assimilation, and CO_2 is given off by the plant because of respiration. By convention, CO_2 assimilation is negative in this part of the light-response curve. As the photon flux increases, photosynthetic CO_2 assimilation increases until it equals CO_2 release by mitochondrial respiration. The point at which CO_2 uptake exactly balances CO_2 release is called the light compensation point.

The photon flux at which different leaves reach the light compensation point varies with species and developmental conditions. One of the more interesting differences is found between plants grown in full sunlight and those grown in the shade. Light compensation points of sun plants range from 10 to 20 μmol m^{-2} s^{-1}; corresponding values for shade plants are 1 to 5 μmol m^{-2} s^{-1}. The values for shade plants are lower because respiration rates in shade plants are very low, so little net photosynthesis suffices to bring the net rates of CO_2 exchange to zero. Low respiratory rates seem to represent a basic adaptation that allows shade plants to survive in light-limited environments. Increasing photon flux above the light compensation point results in a proportional increase in photosynthetic rate, yielding a linear relationship between photon flux and photosynthetic rate. Such a linear relationship comes about because photosynthesis is *light limited* at those levels of incident light, so more light stimulates more photosynthesis.

In this linear portion of the curve, the slope of the line reveals the maximum quantum yield of photosynthesis for the leaf. Recall that quantum yield is the relation between a given light-dependent product (in this case CO_2 assimilation) and the number of absorbed photons. Quantum yields vary from 0, where none of the light energy is used in photosynthesis, to 1, where all the absorbed light is used. The quantum yield of photochemistry is about 0.95, and the quantum yield of oxygen evolution by isolated chloroplasts is about 0.1 (10 photons per molecule of O_2). In the intact leaf, measured quantum yields for CO_2 fixation vary between 0.04 and 0.06. Healthy leaves from many species of C_3 plants, kept under low O_2 concentrations that inhibit photorespiration, usually show a quantum yield of 0.1. In normal air, the quantum yield of C_3 plants is lower, typically 0.05.

Quantum yield varies with temperature and CO_2 concentration because of their effect on the ratio of the carboxylase and oxygenase reactions of rubisco. Below 30°C, quantum yields of C_3 plants are generally higher than those of C_4 plants; above 30°C, the situation is usually reversed. Despite their different growth habitats, sun and shade plants show similar quantum yields. At higher photon fluxes, the photosynthetic response to light starts to level off and reaches *saturation*. Once the saturation point is reached, further increases in photon flux no longer affect photosynthetic rates, indicating that factors other than incident light, such as electron transport rate, rubisco activity, or the metabolism of triose phosphates, have become limiting to photosynthesis.

After the saturation point, photosynthesis is commonly referred to as CO_2 *limited*, reflecting the inability of the Calvin cycle enzymes to keep pace with the absorbed light energy. Light saturation levels for shade plants are substantially lower than those for sun plants. These levels usually reflect the maximal photon flux to which the leaf was exposed during growth. The light-response curve of most leaves saturates between 500 and 1000 μmol m^{-2} s^{-1}, photon fluxes well below full sunlight (which is about 2000 μmol m^{-2} s^{-1}). Although individual leaves are rarely able to utilize full sunlight, whole plants usually consist of many leaves that shade each other.

For example, only a small fraction of a tree's leaves are exposed to full sun at any given time of the day. The rest of the leaves receive subsaturating photon fluxes in the form of small patches of light that pass through gaps in the leaf canopy or in the form of light transmitted through other leaves. Because the photosynthetic response of the intact plant is the sum of the photosynthetic activity of all the leaves, only rarely is photosynthesis saturated at the level of the whole plant.

Light-response curves of individual trees and of the forest canopy show that photosynthetic rate increases with photon flux and photosynthesis usually does not saturate, even in full sunlight. Along these lines, crop productivity is related to the total amount of light received during the growing season, and given enough water and nutrients, the more light a crop receives, the higher the biomass.

Leaves Must Dissipate Excess Light Energy

When exposed to excess light, leaves must dissipate the surplus absorbed light energy so that it does not harm the photosynthetic apparatus. There are several routes for energy dissipation involving *non-photochemical quenching*, which is the quenching of chlorophyll fluorescence by mechanisms other than photochemistry. The most important example involves the transfer of absorbed light energy away from electron transport towards heat production. Although the molecular mechanisms are not yet fully understood, the xanthophyll cycle appears to be an important avenue for dissipation of excess light energy.

The Xanthophyll Cycle

RThe xanthophyll cycle, which comprises the three carotenoids violaxanthin, antheraxanthin, and zeaxanthin, is involved in the dissipation of excess light energy in the leaf. Under high light, violaxanthin is converted to antheraxanthin and then to zeaxanthin. Note that the two aromatic rings of violaxanthin have a bound oxygen atom in them, antheraxanthin has one, and zeaxanthin has none. Experiments have shown that zeaxanthin is the most effective of the three xanthophylls in heat dissipation, and antheraxanthin is only half as effective. Whereas the levels of antheraxanthin remain relatively constant throughout the day, the zeaxanthin content increases at high irradiances and decreases at low irradiances. In leaves growing under full sunlight, zeaxanthin and antheraxanthin can make up 60 per cent of the total xanthophyll cycle pool at maximal irradiance levels attained at midday.

In these conditions a substantial amount of excess light energy absorbed by the thylakoid membranes can be dissipated as heat, thus preventing damage to the photosynthetic machinery of the chloroplast. The fraction of light energy that is dissipated depends on irradiance, species, growth conditions, nutrient status, and ambient temperature.

The Xanthophyll Cycle in Sun and Shade Leaves

Leaves that grow in full sunlight contain a substantially larger xanthophyll pool than shade leaves, so they can dissipate higher amounts of excess light energy. Nevertheless, the xanthophyll cycle also operates in plants that grow in the low light of the forest understory, where they are only occasionally exposed to high light when sunlight passes through gaps in the overlying leaf canopy, forming sunflecks. Exposure to one sunfleck results in the conversion of much of the violaxanthin in the leaf to zeaxanthin.

In contrast to typical leaves, in which violaxanthin levels increase again when irradiances drop, the zeaxanthin formed in shade leaves of the forest understory is retained and protects the leaf against exposure to subsequent sunflecks. The xanthophyll cycle is also found in species such as conifers, the leaves of which remain green during winter, when photosynthetic rates are very low yet light absorption remains high. Contrary to the diurnal cycling of the xanthophyll pool observed in the summer, zeaxanthin levels remain high all day during the winter. Presumably this mechanism maximizes dissipation of light energy, thereby protecting the leaves against photooxidation during winter.

In addition to protecting the photosynthetic system against high light, the xanthophyll cycle may help protect against high temperatures. Chloroplasts are more tolerant of heat when they accumulate zeaxanthin. Thus, plants may employ more than one biochemical mechanism to guard against the deleterious effect of excess heat.

Leaves Must Dissipate Vast Quantities of Heat

The heat load on a leaf exposed to full sunlight is very high. In fact, a leaf with an effective thickness of water of 300 ìm would warm up by 100°C every minute if all available solar energy were absorbed and no heat were lost.

However, this enormous heat load is dissipated by the emission of longwave radiation, by sensible (*i.e.*, perceptible) heat loss, and by evaporative (or latent) heat loss:

- Air circulation around the leaf removes heat from the leaf surfaces if the temperature of the leaf is higher than that of the air; this phenomenon is called sensible heat loss.
- Evaporative heat loss occurs because the evaporation of water requires energy. Thus as water evaporates from a leaf, it withdraws heat from the leaf and cools it. The human body is cooled by the same principle, through perspiration.

Sensible heat loss and evaporative heat loss are the most important processes in the regulation of leaf temperature, and the ratio of the two is called the Bowen ratio:

$$\text{Bowen Ratio} = \frac{\text{Sensible heat loss}}{\text{Evaporative heat loss}}$$

In well-watered crops, transpiration, and hence water evaporation from the leaf, is high, so the Bowen ratio is low. On the other hand, when evaporative cooling is limited, the Bowen ratio is large. For example, in some cacti, stomata closure prevents evaporative cooling; all the heat is dissipated by sensible heat loss, and the Bowen ratio is infinite. Plants with very high Bowen ratios conserve water but have to endure very high leaf temperatures in order to maintain a sufficient temperature gradient between the leaf and the air. Slow growth is usually correlated with these adaptations.

Isoprene Synthesis Helps Leaves Cope with Heat

We have seen how the xanthophyll cycle can protect against high light, but how do chloroplasts cope with the high leaf temperatures that usually accompany high light? Isoprene synthesis appears to confer stability to photosynthetic membranes at high light and temperatures. Many plants, including American oak (*Quercus* sp.), aspen (*Populus* sp.), and kudzu (*Pueraria lobata*) emit gaseous five-carbon molecules such as isoprene. On a global scale, these emissions amount to 5×10^{14} g released to the atmosphere each year. These gaseous hydrocarbons are responsible for the pine scent (α- and β-pinene) in coniferous forests and can form a blue haze above forests on hot days. Because isoprene and related hydrocarbons play an important role in atmospheric chemistry, they have attracted much attention from atmospheric scientists. Isoprene emission from leaves can constitute a significant fraction of the carbon assimilated in photosynthesis. For example, up to 2 per cent of the carbon fixed by photosynthesis in aspen and oak leaves at 30°C is released as isoprene. Sun leaves synthesize more isoprene than shade leaves, and synthesis is proportional to leaf temperature and water stress.

Evidence that isoprene confers stability to photosynthetic membranes under high temperatures comes from three types of experimental results:

- Whereas preventing isoprene emission with an inhibitor increases susceptibility to damage by heat, adding isoprene to plants that do not produce isoprene confers heat stability.
- Mutant plants unable to emit isoprene are more easily damaged by high temperatures than are wild-type plants.
- Isoprene is rapidly synthesized enzymatically in response to elevated leaf temperatures.

Absorption of Too Much Light Can Lead to Photoinhibition

When leaves are exposed to more light than they can utilize, the reaction centre of PSII is inactivated and damaged, in a phenomenon called photoinhibition. The characteristics of photoinhibition in the intact leaf depend on the amount of light to which the plant is exposed, and two types of photoinhibition are identified: dynamic photoinhibition and chronic photoinhibition. Under moderate excess light, dynamic photoinhibition is observed. Quantum efficiency decreases, but the maximum photosynthetic rate remains unchanged.

Dynamic photoinhibition is caused by the diversion of absorbed light energy towards heat dissipation—hence the decrease in quantum efficiency. This decrease is often temporary, and quantum efficiency can return to its initial higher value when photon flux decreases below saturation levels. Chronic photoinhibition results from exposure to high levels of excess light that damage the photosynthetic system and decrease both quantum efficiency and maximum photosynthetic rate.

Chronic photoinhibition is associated with damage and replacement of the D1 protein from the reaction centre of PSII. In contrast to dynamic photoinhibition, these effects are relatively long-lasting, persisting for weeks or months. Early researchers of photoinhibition interpreted all decreases in quantum efficiency as damage to the photosynthetic apparatus. It is now recognized that short-term decreases in quantum efficiency seem to reflect protective mechanisms, whereas chronic photoinhibition represents actual damage to the chloroplast resulting from excess light, or a failure of the protective mechanisms.

How significant is photoinhibition in nature? Dynamic photoinhibition appears to occur normally at midday, when leaves are exposed to maximum amounts of light and there is a corresponding reduction in carbon fixation. Photoinhibition is more pronounced at low temperatures, and it becomes chronic under more extreme climatic conditions.

Studies of natural willow populations, and crops of *Brassica napus* (oilseed rape) and *Zea mays* (maize), have shown that the cumulative effects of a daily depression in photosynthetic rates caused by photoinhibition decrease biomass by 10 per cent at the end of the growing season. This may not seem a particularly large effect, but it could be significant in natural plant populations competing for limited resources—conditions under which any reduction in carbon allocated to reproduction can adversely affect reproductive success and survival.

STAGES OF PHOTOSYNTHESIS

Photosynthesis is a two stage process. The first process is the Light Dependent Process (Light Reactions), requires the direct energy of light to make energy carrier molecules that are used in the second process.

The Light Independent Process (or Dark Reactions) occurs when the products of the Light Reaction are used to form C–C covalent bonds of carbohydrates. The Dark Reactions can usually occur in the dark, if the energy carriers from the light process are present. Recent evidence suggests that a major enzyme of the Dark Reaction is indirectly stimulated by light, thus the term Dark Reaction is somewhat of a misnomer. The Light Reactions occur in the grana and the Dark Reactions take place in the stroma of the chloroplasts.

LIGHT REACTIONS

In the Light Dependent Processes (Light Reactions) light strikes chlorophyll a in such a way as to excite electrons to a higher energy state. In a series of reactions the energy is converted (along an electron transport process) into ATP and NADPH. Water is split in the process, releasing oxygen as a by–product of the reaction. The ATP and NADPH are used to make C–C bonds in the Light Independent Process (Dark Reactions). In the Light Independent Process, carbon dioxide from the atmosphere (or water for aquatic/marine organisms) is captured and modified by the addition of Hydrogen to form carbohydrates (general formula of carbohydrates is $[CH_2O]_n$). The incorporation of carbon dioxide into organic compounds is known as carbon fixation.

The energy for this comes from the first phase of the photosynthetic process. Living systems cannot directly utilize light energy, but can, through a complicated series of reactions, convert it into C–C bond energy that can be released by glycolysis and other metabolic processes. Photosystems are arrangements of chlorophyll and other pigments packed into thylakoids. Many Prokaryotes have only one photosystem, Photosystem II (so numbered because, while it was most likely the first to evolve, it was the second one discovered). Eukaryotes have Photosystem II plus Photosystem I. Photosystem I uses chlorophyll a, in the form referred to as P700. Photosystem II uses a form of chlorophyll a known as P680. Both "active" forms of chlorophyll a function in photosynthesis due to their association with proteins in the thylakoid membrane.

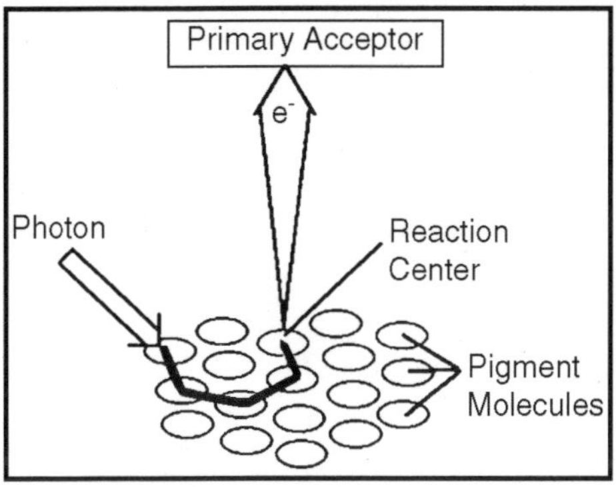

Fig. Action of a Photosystem

Photophosphorylation is the process of converting energy from a light–excited electron into the pyrophosphate bond of an ADP molecule. This occurs when the electrons from water are excited by the light in the presence of P680. The energy transfer is similar to the chemiosmotic electron transport occurring in the mitochondria. Light energy causes the removal of an electron from a molecule of P680 that is part of Photosystem II. The P680 requires an electron, which is taken from a water molecule, breaking the water into H^+ ions and O^{-2} ions. These O^{-2} ions combine to form the diatomic O_2 that is released. The electron is "boosted" to a higher energy state and attached to a primary electron acceptor, which begins a series of redox reactions, passing the electron through a series of electron carriers, eventually attaching it to a molecule in Photosystem I. Light acts on a molecule of P700 in Photosystem I, causing an electron to be "boosted" to a still higher potential.

The electron is attached to a different primary electron acceptor (that is a different molecule from the one associated with Photosystem II). The electron is passed again through a series of redox reactions, eventually being attached to $NADP^+$ and H^+ to form NADPH, an energy carrier needed in the Light Independent Reaction. The electron from Photosystem II replaces the excited electron in the P700 molecule.

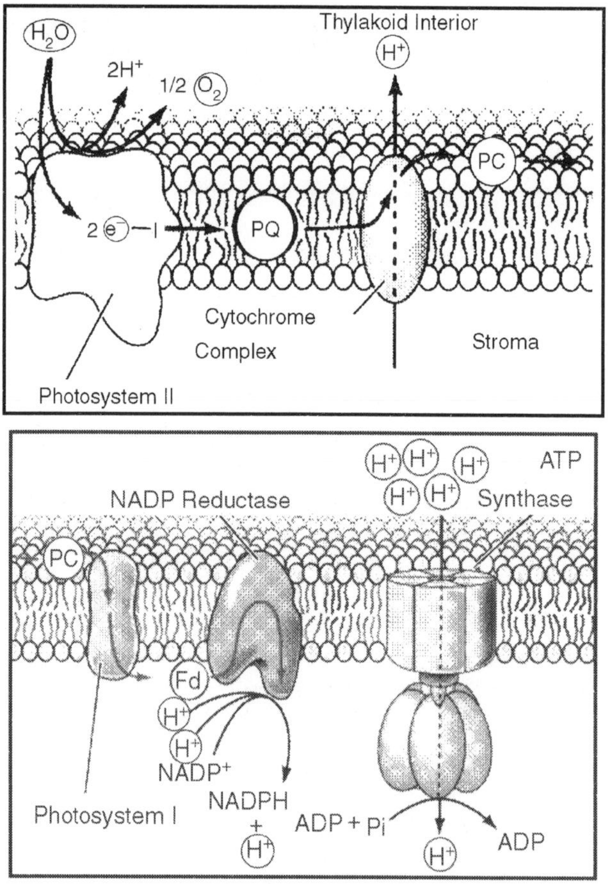

Fig. Chemiosmosis as it Operates in PhotophosphorylationWithin a Chloroplast.

There is thus a continuous flow of electrons from water to NADPH. This energy is used in Carbon Fixation. Cyclic Electron Flow occurs in some eukaryotes and primitive photosynthetic bacteria. No NADPH is produced, only ATP. This occurs when cells may require additional ATP, or when there is no $NADP^+$ to reduce to NADPH. In Photosystem II, the pumping to H ions into the thylakoid and the conversion of ADP + P into ATP is driven by electron gradients established in the thylakoid membrane.

The above diagrams present the "old" view of photophosphorylation. We now know where the process occurs in the chloroplast, and can link that to chemiosmotic synthesis of ATP.

Halobacteria, which grow in extremely salty water, are facultative aerobes, they can grow when oxygen is absent. Purple pigments, known as retinal (a pigment also found in the human eye) act similar to chlorophyll. The complex of retinal and membrane proteins is known as bacteriorhodopsin, which generates electrons which establish a proton gradient that powers an ADP–ATP pump, generating ATP from sunlight without chlorophyll. This supports the theory that chemiosmotic processes are universal in their ability to generate ATP.

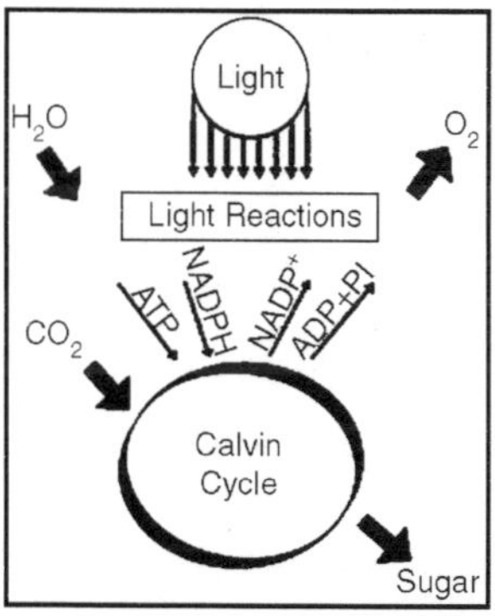

Fig. Photosynthesis Changes the Energy from the Sun Into Chemical Energy, Splits Water to Liberate O_2, and Fixes CO_2 Into Sugar.

DARK REACTION

Carbon–Fixing Reactions are also known as the Dark Reactions (or Light Independent Reactions). Carbon dioxide enters single–celled and aquatic autotrophs through no specialized structures, diffusing into the cells. Land plants must guard against drying out (desiccation) and so have evolved specialized structures known as stomata to allow gas to enter and leave the leaf. The Calvin Cycle occurs in the stroma of chloroplasts (where would it occur in a prokaryote?). Carbon dioxide is captured by the chemical ribulose biphosphate (RuBP). RuBP is a 5–C chemical. Six molecules of carbon dioxide enter the Calvin Cycle, eventually producing one molecule of glucose.

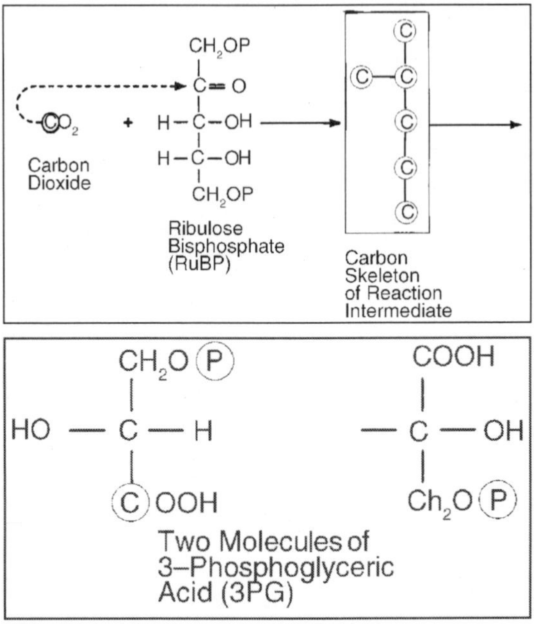

Fig. The First Steps in the Calvin cycle.

The first stable product of the Calvin Cycle is phosphoglycerate (PGA), a 3–C chemical.

The energy from ATP and NADPH energy carriers generated by the photosystems is used to attach phosphates to (phosphorylate) the PGA. Eventually there are 12 molecules of glyceraldehyde phosphate (also known as phosphoglyceraldehyde or PGAL, a 3–C), two of which are removed from the cycle to make a glucose.

The remaining PGAL molecules are converted by ATP energy to reform 6 RuBP molecules, and thus start the cycle again. Remember the complexity of life, each reaction in this process, as in Kreb's Cycle, is catalyzed by a different reaction–specific enzyme.

C–4 Pathway

Some plants have developed a preliminary step to the Calvin Cycle (which is also referred to as a C–3 pathway), this preamble step is known as C–4. While most C–fixation begins with RuBP, C–4 begins with a new molecule, phosphoen-olpyruvate (PEP), a 3–C chemical that is converted into oxaloacetic acid (OAA, a 4–C chemical) when carbon dioxide is combined with PEP.

The OAA is converted to Malic Acid and then transported from the mesophyll cell into the bundle–sheath cell, where OAA is broken down into PEP plus carbon dioxide. The carbon dioxide then enters the Calvin Cycle, with PEP returning to the mesophyll cell. The resulting sugars are now adjacent to the leaf veins and can readily be transported throughout the plant.

The capture of carbon dioxide by PEP is mediated by the enzyme PEP carboxylase, which has a stronger affinity for carbon dioxide than does RuBP carboxylase When carbon dioxide levels decline below the threshold for RuBP carboxylase, RuBP is catalyzed with oxygen instead of carbon dioxide. The product of that reaction forms glycolic acid, a chemical that can be broken down by photorespiration, producing neither NADH nor ATP, in effect dismantling the Calvin Cycle. C–4 plants, which often grow close together, have had to adjust to decreased levels of carbon dioxide by artificially raising the carbon dioxide concentration in certain cells to prevent photorespiration.

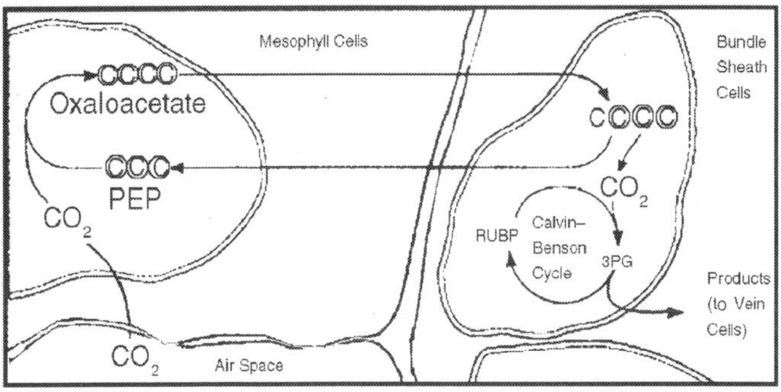

Fig. C–4 Photosynthsis Involves the Separation of Carbon Fixation and Carbohydrate Systhesis in Space and Time.

C–4 plants evolved in the tropics and are adapted to higher temperatures than are the C–3 plants found at higher latitudes. Common C–4 plants include crabgrass, corn, and sugar cane. Note that OAA and Malic Acid also have functions in other processes, thus the chemicals would have been present in all plants, leading scientists to hypothesize that C–4 mechanisms evolved several times independently in response to a similar environmental condition, a type of evolution known as convergent evolution.

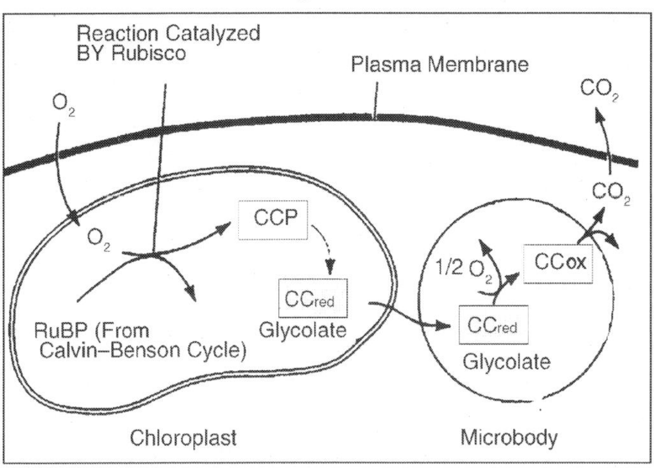

Fig. Photorespiration.

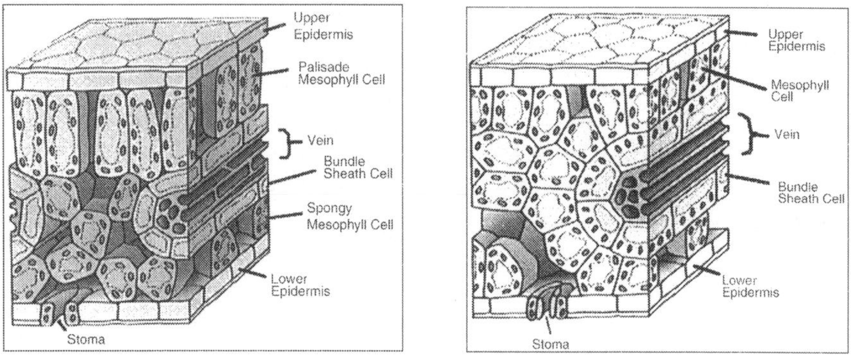

Fig. Leaf Anatomy of a C3 (top) and C4 (Bottom) Plant.

We can see anatomical differences between C3 and C4 leaves.

Carbon Cycle

Plants may be viewed as carbon sinks, removing carbon dioxide from the atmosphere and oceans by fixing it into organic chemicals.

Plants also produce some carbon dioxide by their respiration, but this is quickly used by photosynthesis. Plants also convert energy from light into chemical energy of C–C covalent bonds. Animals are carbon dioxide producers that derive their energy from carbohydrates and other chemicals produced by plants by the process of photosynthesis.

The balance between the plant carbon dioxide removal and animal carbon dioxide generation is equalized also by the formation of carbonates in the oceans. This removes excess carbon dioxide from the air and water (both of which are in equilibrium with regard to carbon dioxide).

Fossil fuels, such as petroleum and coal, as well as more recent fuels such as peat and wood generate carbon dioxide when burned. Fossil fuels are formed ultimately by organic processes, and represent also a tremendous carbon sink. Human activity has greatly increased the concentration of carbon dioxide in air.

This increase has led to global warming, an increase in temperatures around the world, the Greenhouse Effect.

The increase in carbon dioxide and other pollutants in the air has also led to acid rain, where water falls through polluted air and chemically combines with carbon dioxide, nitrous oxides, and sulfur oxides, producing rainfall with pH as low as 4.

This results in fish kills and changes in soil pH which can alter the natural vegetation and uses of the land.

The Global Warming problem can lead to melting of the ice caps in Greenland and Antarctica, raising sea–level as much as 120 meters. Changes in sea–level and temperature would affect climate changes, altering belts of grain production and rainfall patterns.

PHOTOSYNTHETIC RESPONSESTO CARBON DIOXIDE

We have discussed how plant growth and leaf anatomy are influenced by light. Now we turn our attention to how CO_2 concentration affects photosynthesis. CO_2 diffuses from the atmosphere into leaves—first through stomata, then through the intercellular air spaces, and ultimately into cells and chloroplasts. In the presence of adequate amounts of light, higher CO_2 concentrations support higher photosynthetic rates. The reverse is also true; that is, low CO_2 concentration can limit the amount of photosynthesis. The concentration of atmospheric CO_2 in recent history, and its availability for carbon-fixing processes. Then we'll consider the limitations that CO_2 places on photosynthesis and the impact of the CO_2-concentrating mechanisms of C_4 plants.

ATMOSPHERIC CO_2 CONCENTRATION KEEPS RISING

Carbon dioxide is a trace gas in the atmosphere, presently accounting for about 0.037 per cent, or 370 parts per million (ppm), of air. The partial pressure of ambient CO_2 (*C*a) varies with atmospheric pressure and is approximately 36 pascals (Pa) at sea level. Water vapor usually accounts for up to 2 per cent of the atmosphere and O_2 for about 20 per cent. The bulk of the atmosphere, nearly 80 per cent, is nitrogen.

The current atmospheric concentration of CO_2 is almost twice the concentration that has prevailed during most of the last 160,00 years, as measured from air bubbles trapped in glacial ice in Antarctica. Except for the last 200 years, CO_2 concentrations during the recent geological past have been low, fluctuating between 180 and 260 ppm. These low concentrations were typical of times extending back to the Cretaceous, when Earth was much warmer and the CO_2 concentration may have been as high as 1200 to 2800 ppm.

The current CO_2 concentration of the atmosphere is increasing by about 1 ppm each year, primarily because of the burning of fossil fuels. Since 1958, when systematic measurements of CO_2 began at Mauna Loa, Hawaii, atmospheric CO_2 concentrations have increased by more than 17 per cent, and by 2020 the atmospheric CO_2 concentration could reach 600 ppm.

The Greenhouse Effect

The consequences of this increase in atmospheric CO_2 are under intense scrutiny by scientists and government agencies, particularly because of predictions that the greenhouse effect is altering the world's climate. The term *greenhouse effect* refers to the resulting warming of Earth's climate, which is caused by the trapping of long-wavelength radiation by the atmosphere. A greenhouse roof transmits visible light, which is absorbed by plants and other surfaces inside the greenhouse. The absorbed light energy is converted to heat, and part of it is re-emitted as long-wavelength radiation.

Because glass transmits long-wavelength radiation very poorly, this radiation cannot leave the greenhouse through the glass roof, and the greenhouse heats up. Certain gases in the atmosphere, particularly CO_2 and methane, play the same role as the glass roof in a greenhouse. The increased CO_2 concentration and temperature associated with the greenhouse effect can influence photosynthesis. At current atmospheric CO_2 concentrations, photosynthesis in C_3 plants is CO_2 limited, but this situation could change as atmospheric CO_2 concentrations continue to rise. Under laboratory conditions, most C_3 plants grow 30 to 60 per cent faster when CO_2 concentration is doubled (to 600–700 ppm), and the growth rate changes depend on nutrient status. In some plants the enhanced growth is only temporary. For many crops, such as tomatoes, lettuce, cucumbers, and roses growing in greenhouses under optimal nutrition, carbon dioxide enrichment in the greenhouse environment results in increased productivity. The photosynthetic performance of C_3 plants under elevated CO_2 is enhanced because photorespiration decreases.

Diffusion of CO_2 to the Chloroplastis Essential to Photosynthesis

For photosynthesis to occur, carbon dioxide must diffuse from the atmosphere into the leaf and into the carboxylation site of rubisco. Because diffusion rates depend on concentration gradients, appropriate gradients are needed to ensure adequate diffusion of CO_2 from the leaf surface to the chloroplast. The cuticle that covers the leaf is nearly impermeable to CO_2, so the main port of entry of CO_2 into the leaf is the stomatal pore. CO_2 diffuses through the pore into the substomatal cavity and into the intercellular air spaces between the mesophyll cells.

This portion of the diffusion path of CO_2 into the chloroplast is a gaseous phase. The remainder of the diffusion path to the chloroplast is a liquid phase, which begins at the water layer that wets the walls of the mesophyll cells and continues through the plasma membrane, the cytosol, and the chloroplast.

Each portion of this diffusion pathway imposes a resistance to CO_2 diffusion, so the supply of CO_2 for photosynthesis meets a series of different points of resistance. An evaluation of the magnitude of each point of resistance is helpful for understanding CO_2 limitations to photosynthesis. Carbon dioxide enters the intercellular air spaces of the leaf through the stomatal pores. From the air spaces it dissolves in the water of wet cell walls and diffuses into the cell and chloroplast. The same path is traveled in the reverse direction by H_2O. The sharing of this pathway by CO_2 and water presents the plant with a functional dilemma. In air of high relative humidity, the diffusion gradient that drives water loss is about 50 times larger than the gradient that drives CO_2 uptake. In drier air, this gradient can be even larger. Therefore, a decrease in stomatal resistance through the opening of stomata facilitates higher CO_2 uptake but is unavoidably accompanied by substantial water loss. The gas phase of CO_2 diffusion into the leaf can be divided into three components—the boundary layer, the stomata, and the intercellular spaces of the leaf—each of which imposes a resistance to CO_2 diffusion.

The boundary layer consists of relatively unstirred air at the leaf surface, and its resistance to diffusion is called the boundary layer resistance. The magnitude of the boundary layer resistance decreases with leaf size and wind speed. The boundary layer resistance to water and CO_2 diffusion is physically related to the boundary layer resistance to sensible heat loss discussed earlier. Smaller leaves have a lower boundary layer resistance to CO_2 and water diffusion, and to sensible heat loss. Leaves of desert plants are usually small, facilitating sensible heat loss. The large leaves often found in the humid Tropics can have large boundary layer resistances, but these leaves can dissipate the radiation heat load by evaporative cooling because of the high transpiration rates made possible by the abundant water supply in these habitats.

After diffusing through the boundary layer, CO_2 enters the leaf through the stomatal pores, which impose the next type of resistance in the diffusion pathway, the stomatal resistance. Under most conditions in nature, in which the air around a leaf is seldom completely still, the boundary layer resistance is much smaller than the stomatal resistance, and the main limitation to CO_2 diffusion is imposed by the stomatal resistance.

There is also a resistance to CO_2 diffusion in the air spaces that separate the substomatal cavity from the walls of the mesophyll cells, called the intercellular air space resistance. This resistance is also usually small—causing a drop of 0.5 Pa or less in partial pressure of CO_2, compared with the 36 Pa outside the leaf. The resistance to CO_2 diffusion of the liquid phase—the liquid phase resistance, also called mesophyll resistance—encompasses diffusion from the intercellular leaf spaces to the carboxylation sites in the chloroplast. This point of resistance to CO_2 diffusion has been calculated as approximately one-tenth of the combined boundary layer resistance and stomatal resistance when the stomata are fully open. This low resistance value can be attributed in part to the large surface area of mesophyll cells exposed to the intercellular air spaces, which can be as much as 10 to 30 times the projected leaf area. In addition, the localization of chloroplasts near the cell periphery minimizes the distance that CO_2 diffuses to carboxylation sites within the chloroplast. The positioning of chloroplasts and the relatively large percentage of intercellular air space (about 20–40 per cent) are special anatomic features that facilitate the internal diffusion and uptake of CO_2 by leaves.

Because the stomatal pores usually impose the largest resistance to CO_2 uptake and water loss in the diffusion pathway, this regulation provides the plant with an effective way to control gas exchange between the leaf and the atmosphere. In experimental measurements of gas exchange from leaves, the boundary layer resistance and the intercellular air space resistance are usually ignored, and the stomatal resistance is used as the single parameter describing the gas phase resistance to CO_2.

Patterns of Light Absorption Generate Gradients of CO_2 Fixation within the Leaf

We have discussed how leaf anatomy is specialized for capturing light and how it also facilitates the internal diffusion of CO_2, but where in the leaf do maximum rates of photosynthesis occur? In most leaves, light is preferentially absorbed at the upper surface, whereas CO_2 enters through the lower surface. Given that light and CO_2 enter from opposing sides of the leaf, does photosynthesis occur uniformly within the leaf tissues, or is there a gradient in photosynthesis across the leaf?

The photosynthetic properties of a leaf are determined by the following:

- Profiles of light absorption across the mesophyll
- Photosynthetic capacity of those tissues
- Internal CO_2 supply

For most leaves, internal CO_2 diffusion is rapid, so limitations on photosynthetic performance within the leaf are imposed by factors other than CO_2 supply. When white light enters the upper surface of a leaf, blue and red photons are preferentially absorbed by chloroplasts near the irradiated surface, owing to the strong absorption bands of chlorophyll in the blue and red regions of the spectrum. Green light, on the other hand, penetrates deeper into the leaf. Compared to blue and red, chlorophyll absorbs poorly in the green, yet green light is very effective in supplying energy for photosynthesis in the tissues within the leaf depleted from blue and red photons.

The capacity of the leaf tissue for photosynthetic CO_2 assimilation depends to a large extent on its rubisco content. In spinach and the faba bean (*Vicia faba*), rubisco content starts out low at the top of the leaf, increases towards the middle, and then decreases again towards the bottom. As a result, the distribution of carbon fixation within the leaf is bell shaped. The spongy mesophyll fixes about 40 per cent of the total carbon in spinach. The functional significance of the rubisco distribution and the profiles of carbon assimilation within leaves is not yet known, although it is likely that photosynthesis profiles vary in leaves with different anatomy and in leaves adapted to different environments.

CO_2 Imposes Limitations on Photosynthesis

Expressing photosynthetic rate as a function of the partial pressure of CO_2 in the intercellular air space (*C*i) within the leaf makes it possible to evaluate limitations to photosynthesis imposed by CO_2 supply. At very low intercellular CO_2 concentrations, photosynthesis is strongly limited by the low CO_2, while respiratory rates are unaffected. As a result, there is a negative balance between CO_2 fixed by photosynthesis and CO_2 produced by respiration, and a net efflux of CO_2 from the plant. Increasing intercellular CO_2 to the concentration at which these two processes balance each other defines the CO_2 compensation point, at which the net efflux of CO_2 from the plant is zero. This concept is analogous to that of the light compensation point discussed earlier in the chapter: *The CO_2 compensation point reflects the balance between photosynthesis and respiration as a function of CO_2 concentration, and the light compensation point reflects that balance as a function of photon flux.*

In C_3 plants, increasing CO_2 above the compensation point stimulates photosynthesis over a wide concentration range. At low to intermediate CO_2 concentrations, photosynthesis is limited by the carboxylation capacity of rubisco. At high CO_2 concentrations, photosynthesis is limited by the capacity of Calvin cycle to regenerate the acceptor molecule ribulose-1,5-bisphosphate, which depends on electron transport rates. By regulating stomatal conductance, most leaves appear to regulate their *C*i (internal partial pressure for CO_2) such that it is intermediate between limitations imposed by carboxylation capacity and the capacity to regenerate ribulose-1,5-bisphosphate. A plot of CO_2 assimilation as a function intercellular partial pressures of CO_2 tells us how photosynthesis is regulated by CO_2, independent of the functioning of stomata.

Inspection of such a plot for C_3 and C_4 plants reveals interesting differences between the two types of carbon metabolism:

- In *C_4 plants*, photosynthetic rates saturate at *C*i values of about 15 Pa, reflecting the effective CO_2-concentrating mechanisms operating in these plants.
- In *C_3 plants*, increasing *C*i levels continue to stimulate photosynthesis over a much broader range.

These results indicate that C_3 plants may benefit more from ongoing increases in atmospheric CO_2 concentrations. In contrast, photosynthesis in C_4 plants is CO_2 saturated at low concentrations, and as a result C_4 plants do not benefit from increases in atmospheric CO_2 concentrations. Plants with C_4 metabolism have a CO_2 compensation point of zero or nearly zero, reflecting their very low levels of photorespiration. This difference between C_3 and C_4 plants is not seen when the experiments are conducted at low oxygen concentrations because oxygenation is also suppressed in C_3 plants.

CO_2-Concentrating MechanismsAffect Photosynthetic Responses of Leaves

Because of the operating CO_2-concentrating mechanisms in C_4 plants, CO_2 concentration at the carboxylation sites within C_4 chloroplasts is often saturating for rubisco activity. As a result, plants with C4 metabolism need less rubisco than C3 plants need to achieve a given rate of photosynthesis, and require less nitrogen to grow. In addition, the CO_2-concentrating mechanism allows the leaf to maintain high photosynthetic rates at lower *Ci* values, which require lower rates of stomatal conductance for a given rate of photosynthesis. Thus, C_4 plants can use water and nitrogen more efficiently than C_3 plants can. On the other hand, the additional energy cost of the concentrating mechanism makes C_4 plants less efficient in their utilization of light. This is probably one of the reasons that most shade-adapted plants are C_3 plants. Many cacti and other succulent plants with CAM metabolism open their stomata at night and close them during the day. The CO_2 taken up during the night is fixed into malate. Because air temperatures are much lower at night than during the day, water loss is low and a significant amount of water is saved relative to the amount of CO_2 fixed.

The main constraint on CAM metabolism is that the capacity to store malic acid is limited, and this limitation restricts the amount of CO_2 uptake. However, many CAM plants can fix CO_2 via the Calvin cycle at the end of the day, when temperature gradients are less extreme. Cladodes (flattened stems) of cacti can survive after detachment from the plant for several months without water. Their stomata are closed all the time, and the CO_2 released by respiration is refixed into malate. This process, which has been called *CAM idling*, allows the plant to survive for prolonged periods of time while losing remarkably little water.

Discrimination of Carbon Isotopes
RevealsDifferent Photosynthetic Pathways

Atmospheric CO_2 contains the naturally occurring carbon isotopes ^{12}C, ^{13}C, and ^{14}C in the proportions 98.9 per cent, 1.1 per cent, and 10–10 per cent, respectively. $^{14}CO_2$ is present in such small quantities that it has no physiological relevance, but $^{13}CO_2$ is different. The chemical properties of $^{13}CO_2$ are identical to those of $^{12}CO_2$, but because of the slight difference in mass (2.3 per cent), most plants assimilate less $^{13}CO_2$ than $^{12}CO_2$. In other words, plants discriminate against the heavier isotope of carbon, and they have smaller ratios of ^{13}C to ^{12}C than are found in atmospheric CO_2. How effective are plants at distinguishing between the two carbon isotopes? Although discrimination against ^{13}C is subtle, the isotope composition of plants reveals a wealth of information.

Carbon isotope composition is measured by use of a mass spectrometer, which yields the following ratio:

$$R = \frac{^{13}CO_2}{^{12}CO_2}$$

The isotope composition of plants, $\delta^{13}C$, is quantified on a per mil (–%) basis:

$$\delta^{13}C\% = \left(\frac{R_{sample}}{R_{standard}} - 1 \right) \times 1000$$

where the standard represents the carbon isotopes contained in a fossil belemnite from the Pee Dee limestone formation of South Carolina. The $\delta^{13}C$ of atmospheric CO_2 has a value of–8 per cent, meaning that there is less ^{13}C in the atmospheric CO_2 than is found in the carbonate of the belemnite standard. What are some typical values for carbon isotope ratios of plants? C_3 plants have a $\delta^{13}C$ of about–28 per cent; C_4 plants have an average value of–14 per cent. Both C_3 and C_4 plants have less ^{13}C than the isotope standard, which means that there has been a discrimination against ^{13}C during the photosynthetic process. Because the per mil calculation involves multiplying by 1000, the actual isotope discrimination is small. Nonetheless, differences in carbon isotope discrimination are easily detectable with mass spectrometers. For example, measuring the $\delta^{13}C$ of table sugar (sucrose) makes it possible to determine if the sucrose came from sugar beet (a C_3 plant) or sugarcane (a C_4 plant). What is the physiological basis for ^{13}C depletion in plants? One reason in both C_3 and C_4 plants is diffusion. CO_2 diffuses from air outside of the leaf to the carboxylation sites within leaves. Because $^{12}CO_2$ is lighter than $^{13}CO_2$, it diffuses slightly faster towards the carboxylation site, creating an effective diffusion discrimination of –4.4 ‰. However, the largest isotope discrimination step is the carboxylation reaction catalyzed by rubisco. Rubisco has an intrinsic discrimination value against ^{13}C of–30 ‰. By contrast, PEP carboxylase, the primary CO_2 fixation enzyme of C_4 plants, has a much smaller isotope discrimination effect (about–2 to–6 ‰).

Thus the inherent difference between the discrimination effects of the two carboxylating enzymes causes the different isotope compositions observed in C_3 and C_4 plants. Other physiological characteristics of plants affect isotope composition. One factor is the partial pressure of CO_2 in the intercellular air spaces of leaves (*C*i). In C_3 plants the potential discrimination by rubisco of–30 per cent is not fully expressed because the availability of CO_2 at the carboxylation site becomes a limiting factor restricting the discrimination by rubisco. More discrimination occurs when *C*i is high, as when stomata are open. Open stomata also facilitate water loss. Thus, lower water use efficiency is correlated with greater discrimination against ^{13}C. Fossil fuels have a $\delta^{13}C$ of about–26 per cent because the carbon in these deposits came from organisms that had a C_3 carbon fixation pathway.

Furthermore, measuring $\delta^{13}C$ in fossil, carbonate-containing soils and fossil teeth makes it possible to determine that C_4 photosynthesis developed and became prevalent relatively recently. CAM plants can have $\delta^{13}C$ values that are intermediate between those of C_3 and C_4 plants. In CAM plants that fix CO_2 at night via PEP carboxylase, $\delta^{13}C$ is similar to that of C_4 plants. However, when some CAM plants are well watered, they switch to C_3 mode by opening their stomata and fixing CO_2 during the day via rubisco.

Under these conditions the isotope composition shifts more towards that of C_3 plants. Thus the $^{13}C/^{12}C$ values of CAM plants reflect how much carbon is fixed via the C_3 pathway versus the C_4 pathway. Plants also fractionate other isotopes, such as $^{18}O/^{16}O$ and $^{15}N/^{14}N$, and the various patterns of isotope enrichment or depletion can be used as indicators of particular metabolic pathways or features.

PHOTOSYNTHETIC MEMBRANES AND ORGANELLES

The proteins that gather light for photosynthesis are embedded within cell membranes. The simplest way these are arranged is in photosynthetic bacteria, where these proteins are held within the plasma membrane. However, this membrane may be tightly folded into cylindrical sheets called thylakoids, or bunched up into round vesicles called intracytoplasmic membranes. These structures can fill most of the interior of a cell, giving the membrane a very large surface area and therefore increasing the amount of light that the bacteria can absorb.

In plants and algae, photosynthesis takes place in organelles called chloroplasts. A typical plant cell contains about 10 to 100 chloroplasts. The chloroplast is enclosed by a membrane. This membrane is composed of a phospholipid inner membrane, a phospholipid outer membrane, and an intermembrane space between them. Within the membrane is an aqueous fluid called the stroma. The stroma contains stacks (grana) of thylakoids, which are the site of photosynthesis. The thylakoids are flattened disks, bounded by a membrane with a lumen or thylakoid space within it. The site of photosynthesis is the thylakoid membrane, which contains integral and peripheral membrane protein complexes, including the pigments that absorb light energy, which form the photosystems. Plants absorb light primarily using the pigment chlorophyll, which is the reason that most plants have a green colour. Besides chlorophyll, plants also use pigments such as carotenes and xanthophylls.

Algae also use chlorophyll, but various other pigments are present as phycocyanin, carotenes, and xanthophylls in green algae, phycoerythrin in red algae (rhodophytes) and fucoxanthin in brown algae and diatoms resulting in a wide variety of colours.

These pigments are embedded in plants and algae in special antenna–proteins. In such proteins all the pigments are ordered to work well together. Such a protein is also called a light–harvesting complex. Although all cells in the green parts of a plant have chloroplasts, most of the energy is captured in the leaves. The cells in the interior tissues of a leaf, called the mesophyll, can contain between 450,000 and 800,000 chloroplasts for every square millimeter of leaf. The surface of the leaf is uniformly coated with a water–resistant waxy cuticle that protects the leaf from excessive evaporation of water and decreases the absorption of ultraviolet or blue light to reduce heating. The transparent epidermis layer allows light to pass through to the palisade mesophyll cells where most of the photosynthesis takes place.

LIGHT REACTIONS

In the light reactions, one molecule of the pigment chlorophyll absorbs one photon and loses one electron. This electron is passed to a modified form of chlorophyll called pheophytin, which passes the electron to a quinone molecule, allowing the start of a flow of electrons down an electron transport chain that leads to the ultimate reduction of NADP to NADPH. In addition, this creates a proton gradient across the chloroplast membrane; its dissipation is used by ATP synthase for the concomitant synthesis of ATP. The chlorophyll molecule regains the lost electron from a water molecule through a process called photolysis, which releases a dioxygen (O_2) molecule. The overall equation for the light–dependent reactions under the conditions of non–cyclic electron flow in green plants is:

$$2\,H_2O + 2\,NADP^+ + 3\,ADP + 3\,P_i + \text{light} \rightarrow 2\,NADPH + 2\,H^+ + 3\,ATP + O_2$$

Not all wavelengths of light can support photosynthesis. The photosynthetic action spectrum depends on the type of accessory pigments present. For example, in green plants, the action spectrum resembles the absorption spectrum for chlorophylls and carotenoids with peaks for violet–blue and red light. In red algae, the action spectrum overlaps with the absorption spectrum of phycobilins for blue–green light, which allows these algae to grow in deeper waters that filter out the longer wavelengths used by green plants. The non–absorbed part of the light spectrum is what gives photosynthetic organisms their colour (*e.g.*, green plants, red algae, purple bacteria) and is the least effective for photosynthesis in the respective organisms.

Z Scheme

In plants, light–dependent reactions occur in the thylakoid membranes of the chloroplasts and use light energy to synthesize ATP and NADPH. The light–dependent reaction has two forms: cyclic and non–cyclic. In the non–cyclic reaction, the photons are captured in the light–harvesting antenna complexes of photosystem II by chlorophyll and other accessory pigments. When a chlorophyll molecule at the core of the photosystem II reaction centre obtains sufficient excitation energy from the adjacent antenna pigments, an electron is transferred to the primary electron–acceptor molecule, pheophytin, through a process called photoinduced charge separation. These electrons are shuttled through an electron transport chain, the so called Z–scheme shown in the diagram, that initially functions to generate a chemiosmotic potential across the membrane. An ATP synthase enzyme uses the chemiosmotic potential to make ATP during photophosphorylation, whereas NADPH is aproduct of the terminal redox reaction in the Z–scheme.

The electron enters a chlorophyll molecule in Photosystem I. The electron is excited due to the light absorbed by the photosystem. A second electron carrier accepts the electron, which again is passed down lowering energies of electron acceptors.

The energy created by the electron acceptors is used to move hydrogen ions across the thylakoid membrane into the lumen. The electron is used to reduce the co–enzyme NADP, which has functions in the light–independent reaction. The cyclic reaction is similar to that of the non–cyclic, but differs in the form that it generates only ATP, and no reduced NADP (NADPH) is created. The cyclic reaction takes place only at photosystem I. Once the electron is displaced from the photosystem, the electron is passed down the electron acceptor molecules and returns back to photosystem I, from where it was emitted, hence the name cyclic reaction.

WATER PHOTOLYSIS

The NADPH is the main reducing agent in chloroplasts, providing a source of energetic electrons to other reactions. Its production leaves chlorophyll with a deficit of electrons (oxidized), which must be obtained from some other reducing agent. The excited electrons lost from chlorophyll in photosystem I are replaced from the electron transport chain by plastocyanin.

However, since photosystem II includes the first steps of the Z–scheme, an external source of electrons is required to reduce its oxidized chlorophyll a molecules. The source of electrons in green–plant and cyanobacterial photosynthesis is water.

Two water molecules are oxidized by four successive charge–separation reactions by photosystem II to yield a molecule of diatomic oxygen and four hydrogen ions; the electron yielded in each step is transferred to a redox–active tyrosine residue that then reduces the photoxidized paired–chlorophyll a species called P680 that serves as the primary (light–driven) electron donor in the photosystem II reaction centre.

The oxidation of water is catalyzed in photosystem II by a redox–active structure that contains four manganese ions and a calcium ion; this oxygen–evolving complex binds two water molecules and stores the four oxidizing equivalents that are required to drive the water–oxidizing reaction. Photosystem II is the only known biological enzyme that carries out this oxidation of water.

The hydrogen ions contribute to the transmembrane chemiosmotic potential that leads to ATP synthesis. Oxygen is a waste product of light–dependent reactions, but the majority of organisms on Earth use oxygen for cellular respiration, including photosynthetic organisms.

LIGHT–INDEPENDENT REACTIONS

The Calvin Cycle

In the Light–independent or dark reactions the enzyme RuBisCO captures CO_2 from the atmosphere and in a process that requires the newly formed NADPH, called the Calvin–Benson Cycle, releases three–carbon sugars, which are later combined to form sucrose and starch. The overall equation for the light–independent reactions in green plants is:

$$3\ CO_2 + 9\ ATP + 6\ NADPH + 6\ H^+ \rightarrow$$
$$C_3H_6O_3\text{–phosphate} + 9\ ADP + 8\ P_i + 6\ NADP^+ + 3\ H_2O$$

To be more specific, carbon fixation produces an intermediate product, which is then converted to the final carbohydrate products. The carbon skeletons produced by photosynthesis are then variously used to form other organic compounds, such as the building material cellulose, as precursors for lipid and amino acid biosynthesis, or as a fuel in cellular respiration. The latter occurs not only in plants but also in animals when the energy from plants gets passed through a food chain.

The fixation or reduction of carbon dioxide is a process in which carbon dioxide combines with a five–carbon sugar, ribulose 1,5–bisphosphate (RuBP), to yield two molecules of a three–carbon compound, glycerate 3–phosphate (GP), also known as 3–phosphoglycerate (PGA).

GP, in the presence of ATP and NADPH from the light–dependent stages, is reduced to glyceraldehyde 3–phosphate (G3P). This product is also referred to as 3–phosphogl-yceraldehyde (PGAL) or even as triose phosphate.

Triose is a 3–carbon sugar. Most (5 out of 6 molecules) of the G3P produced is used to regenerate RuBP so the process can continue.

The 1 out of 6 molecules of the triose phosphates not "recycled" often condense to form hexose phosphates, which ultimately yield sucrose, starch and cellulose. The sugars produced during carbon metabolism yield carbon skeletons that can be used for other metabolic reactions like the production of amino acids and lipids.

Carbon Concentrating Mechanisms

On Land

In hot and dry conditions, plants close their stomata to prevent the loss of water. Under these conditions, CO_2 will decrease, and oxygen gas, produced by the light reactions of photosynthesis, will decrease in the stem, not leaves, causing an increase of photorespiration by the oxygenase activity of ribulose–1,5–bisphosphate carboxylase/oxygenase and decrease in carbon fixation. Some plants have evolved mechanisms to increase the CO_2 concentration in the leaves under these conditions. C_4 plants chemically fix carbon dioxide in the cells of the mesophyll by adding it to the three–carbon molecule phosphoenolpyruvate (PEP), a reaction catalyzed by an enzyme called PEP carboxylase, creating the four–carbon organic acid oxaloacetic acid.

Oxaloacetic acid or malate synthesized by this process is then translocated to specialized bundle sheath cells where the enzyme RuBisCO and other Calvin cycle enzymes are located, and where CO_2 released by decarboxylation of the four–carbon acids is then fixed by RuBisCO activity to the three–carbon sugar 3–phosphoglyceric acids.

The physical separation of RuBisCO from the oxygen–generating light reactions reduces photorespiration and increases CO_2 fixation and, thus, photosynthetic capacity of the leaf. C_4 plants can produce more sugar than C_3 plants in conditions of high light and temperature.

Many important crop plants are C_4 plants, including maize, sorghum, sugarcane, and millet. Plants that do not use PEP–carboxylase in carbon fixation are called C_3 plants because the primary carboxylation reaction, catalyzed by RuBisCO, produces the three–carbon sugar 3–phosphoglyceric acids directly in the Calvin–Benson cycle. Over 90 per cent of plants use C_3 carbon fixation, compared to 3 per cent that use C_4 carbon fixation. Xerophytes, such as cacti and most succulents, also use PEP carboxylase to capture carbon dioxide in a process called Crassulacean acid metabolism (CAM).

In contrast to C_4 metabolism, which physically separates the CO_2 fixation to PEP from the Calvin cycle, CAM temporally separates these two processes. CAM plants have a different leaf anatomy from C_3 plants, and fix the CO_2 at night, when their stomata are open.

CAM plants store the CO_2 mostly in the form of malic acid via carboxylation of phosphoenolpyruvate to oxaloacetate, which is then reduced to malate.

Decarboxylation of malate during the day releases CO_2 inside the leaves, thus allowing carbon fixation to 3–phosphoglycerate by RuBisCO. Sixteen thousand species of plants use CAM.

In Water

Cyanobacteria possess carboxysomes, which increase the concentration of CO_2 around RuBisCO to increase the rate of photosynthesis. This operates by carbonic anhydrase, producing hydrocarbonate ions (HCO_3^-), which are then pumped into the carboxysome, before being processed by a different carbonic anhydrase to produce CO_2. Pyrenoids in algae and hornworts also act to concentrate CO_2 around rubisco.

ORDER AND KINETICS

Table. The Overall Process of Photosynthesis takes place in Four Stages.

Stage	Description	Time Scale
1	Energy transfer in antenna chlorophyll	Femtosecond to
	(thylakoid membranes)	picosecond
2	Transfer of electrons in photochemical	Picosecond to
	reactions (thylakoid membranes)	nanosecond
3	Electron transport chain and ATP	Microsecond to
	synthesis (thylakoid membranes)	millisecond
4	Carbon fixation and export of stable products	Millisecond to second

Efficiency

Plants usually convert light into chemical energy with a photosynthetic efficiency of 3–6 per cent. Actual plants' photosynthetic efficiency varies with the frequency of the light being converted, light intensity, temperature and proportion of carbon dioxide in the atmosphere, and can vary from 0.1 per cent to 8 per cent. By comparison, solar panels convert light into electric energy at an efficiency of approximately 6–20 per cent for mass–produced panels, and up to 41 per cent in a research laboratory.

Evolution

Early photosynthetic systems, such as those from green and purple sulfur and green and purple non-sulfur bacteria, are thought to have been anoxygenic, using various molecules as electron donors. Green and purple sulfur bacteria are thought to have used hydrogen and sulfur as an electron donor. Green non-sulfur bacteria used various amino and other organic acids. Purple non-sulfur bacteria used a variety of non-specific organic molecules. The use of these molecules is consistent with the geological evidence that the atmosphere was highly reduced at that time. Fossils of what are thought to be filamentous photosynthetic organisms have been dated at 3.4 billion years old.

The main source of oxygen in the atmosphere is oxygenic photosynthesis, and its first appearance is sometimes referred to as the oxygen catastrophe. Geological evidence suggests that oxygenic photosynthesis, such as that in cyanobacteria, became important during the Paleoproterozoic era around 2 billion years ago. Modern photosynthesis in plants and most photosynthetic prokaryotes is oxygenic. Oxygenic photosynthesis uses water as an electron donor, which is oxidized to molecular oxygen (O_2) in the photosynthetic reaction centre.

SYMBIOSIS AND THE ORIGIN OF CHLOROPLASTS

Several groups of animals have formed symbiotic relationships with photosynthetic algae. These are most common in corals, sponges and sea anemones. It is presumed that this is due to the particularly simple body plans and large surface areas of these animals compared to their volumes. In addition, a few marine mollusks Elysia viridis and Elysia chlorotica also maintain a symbiotic relationship with chloroplasts they capture from the algae in their diet and then store in their bodies. This allows the molluscs to survive solely by photosynthesis for several months at a time. Some of the genes from the plant cell nucleus have even been transferred to the slugs, so that the chloroplasts can be supplied with proteins that they need to survive.

An even closer form of symbiosis may explain the origin of chloroplasts. Chloroplasts have many similarities with photosynthetic bacteria, including a circular chromosome, prokaryotic–type ribosomes, and similar proteins in the photosynthetic reaction centre. The endosymbiotic theory suggests that photosynthetic bacteria were acquired (by endocytosis) by early eukaryotic cells to form the first plant cells. Therefore, chloroplasts may be photosynthetic bacteria that adapted to life inside plant cells. Like mitochondria, chloroplasts still possess their own DNA, separate from the nuclear DNA of their plant host cells and the genes in this chloroplast DNA resemble those in cyanobacteria. DNA in chloroplasts codes for redox proteins such as photosynthetic reaction centers. The CoRR Hypothesis proposes that this Co–location is required for Redox Regulation.

Cyanobacteria and the Evolution of Photosynthesis

The biochemical capacity to use water as the source for electrons in photosynthesis evolved once, in a common ancestor of extant cyanobacteria. The geological record indicates that this transforming event took place early in Earth's history, at least 2450–2320 million years ago (Ma), and, it is speculated, much earlier. Available evidence from geobiological studies of Archean (>2500 Ma) sedimentary rocks indicates that life existed 3500 Ma, but the question of when oxygenic photosynthesis evolved is still unanswered. A clear paleontological window on cyanobacterial evolution opened about 2000 Ma, revealing an already–diverse biota of blue–greens. Cyanobacteria remained principal primary producers throughout the Proterozoic Eon (2500–543 Ma), in part because the redox structure of the oceans favoured photoautotrophs capable of nitrogen fixation.

Green algae joined blue–greens as major primary producers on continental shelves near the end of the Proterozoic, but only with the Mesozoic (251–65 Ma) radiations of dinoflagellates, coccolithophorids, and diatoms did primary production in marine shelf waters take modern form. Cyanobacteria remain critical to marine ecosystems as primary producers in oceanic gyres, as agents of biological nitrogen fixation, and, in modified form, as the plastids of marine algae. A 2010 study by researchers at Tel Aviv University discovered that the Oriental hornet (Vespa orientalis) converts sunlight into electric power using a pigment called xanthopterin. This is the first scientific evidence of a member of the animal kingdom engaging in photosynthesis. Although some of the steps in photosynthesis are still not completely understood, the overall photosynthetic equation has been known since the 19th century. Jan van Helmont began the research of the process in the mid–17th century when he carefully measured the mass of the soil used by a plant and the mass of the plant as it grew. After noticing that the soil mass changed very little, he hypothesized that the mass of the growing plant must come from the water, the only substance he added to the potted plant. His hypothesis was partially accurate—much of the gained mass also comes from carbon dioxide as well as water. However, this was a signaling point to the idea that the bulk of a plant's biomass comes from the inputs of photosynthesis, not the soil itself.

Joseph Priestley, a chemist and minister, discovered that, when he isolated a volume of air under an inverted jar, and burned a candle in it, the candle would burn out very quickly, much before it ran out of wax. He further discovered that a mouse could similarly "injure" air. He then showed that the air that had been "injured" by the candle and the mouse could be restored by a plant. In 1778, Jan Ingenhousz, court physician to the Austrian Empress, repeated Priestley's experiments. He discovered that it was the influence of sunlight on the plant that could cause it to revive a mouse in a matter of hours.

In 1796, Jean Senebier, a Swiss pastor, botanist, and naturalist, demonstrated that green plants consume carbon dioxide and release oxygen under the influence of light. Soon afterwards, Nicolas–Théodore de Saussure showed that the increase in mass of the plant as it grows could not be due only to uptake of CO_2 but also to the incorporation of water. Thus, the basic reaction by which photosynthesis is used to produce food (such as glucose) was outlined. Cornelis Van Niel made key discoveries explaining the chemistry of photosynthesis. By studying purple sulfur bacteria and green bacteria he was the first scientist to demonstrate that photosynthesis is a light–dependent redox reaction, in which hydrogen reduces carbon dioxide.

Robert Emerson discovered two light reactions by testing plant productivity using different wavelengths of light. With the red alone, the light reactions were suppressed. When blue and red were combined, the output was much more substantial. Thus, there were two photosystems, one aborbing up to 600 nm wavelengths, the other up to 700. The former is known as PSII, the latter is PSI. PSI contains only chlorophyll a, PSII contains primarily chlorophyll a with most of the available chlorophyll b, among other pigments. Further experiments to prove that the oxygen developed during the photosynthesis of green plants came from water, were performed by Robert Hill in 1937 and 1939. He showed that isolated chloroplasts give off oxygen in the presence of unnatural reducing agents like iron oxalate, ferricyanide or benzoquinone after exposure to light.

The Hill reaction is as follows:

$$2\ H_2O + 2\ A + (\text{light, chloroplasts}) \rightarrow 2\ AH_2 + O_2$$

where A is the electron acceptor. Therefore, in light, the electron acceptor is reduced and oxygen is evolved. Cyt b_6, now known as a plastoquinone, is one electron acceptor. Samuel Ruben and Martin Kamen used radioactive isotopes to determine that the oxygen liberated in photosynthesis came from the water.

Melvin Calvin and Andrew Benson, along with James Bassham, elucidated the path of carbon assimilation (the photosynthetic carbon reduction cycle) in plants. The carbon reduction cycle is known as the Calvin cycle, which ignores the contribution of Bassham and Benson.

Many scientists refer to the cycle as the Calvin–Benson Cycle, Benson–Calvin, and some even call it the Calvin–Benson–Bassham (or CBB) Cycle. Nobel Prize–winning scientist Rudolph A. Marcus was able to discover the function and significance of the electron transport chain. Otto Heinrich Warburg and Dean Burk discovered the I–quantum photosynthesis reaction that splits the CO2, activated by the respiration.

Factors

There are three main factors affecting photosynthesis and several corollary factors. The three main are:

- Light irradiance and wavelength
- Carbon dioxide concentration
- Temperature.

Light Intensity (Irradiance),Wavelength and Temperature

In the early 20th century, Frederick Frost Blackman along with Albert Einstein investigated the effects of light intensity (irradiance) and temperature on the rate of carbon assimilation.

- At constant temperature, the rate of carbon assimilation varies with irradiance, initially increasing as the irradiance increases. However, at higher irradiance, this relationship no longer holds and the rate of carbon assimilation reaches a plateau.
- At constant irradiance, the rate of carbon assimilation increases as the temperature is increased over a limited range. This effect is seen only at high irradiance levels. At low irradiance, increasing the temperature has little influence on the rate of carbon assimilation.

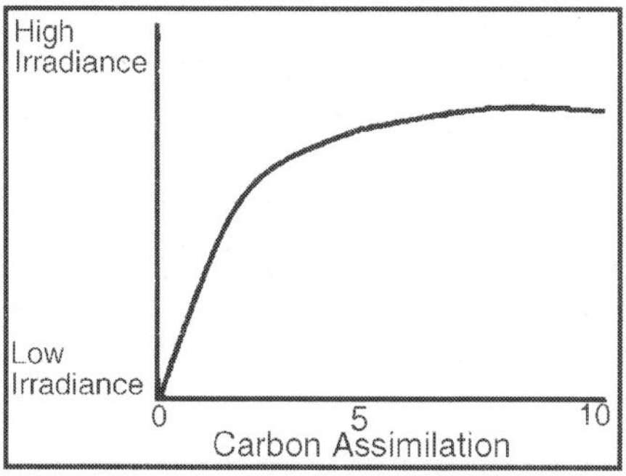

Fig. Carbon Assimilation at a Constant Temperature.

These two experiments illustrate vital points: First, from research it is known that, in general, photochemical reactions are not affected by temperature.

However, these experiments clearly show that temperature affects the rate of carbon assimilation, so there must be two sets of reactions in the full process of carbon assimilation. These are, of course, the light–dependent 'photochemical' stage and the light–independent, temperature–dependent stage. Second, Blackman's experiments illustrate the concept of limiting factors. Another limiting factor is the wavelength of light. Cyanobacteria, which reside several meters underwater, cannot receive the correct wavelengths required to cause photoinduced charge separation in conventional photosynthetic pigments. To combat this problem, a series of proteins with different pigments surround the reaction centre.This unit is called a phycobilisome.

Carbon Dioxide Levels and Photorespiration

As carbon dioxide concentrations rise, the rate at which sugars are made by the light–independent reactions increases until limited by other factors. RuBisCO, the enzyme that captures carbon dioxide in the light–independent reactions, has a binding affinity for both carbon dioxide and oxygen. When the concentration of carbon dioxide is high, RuBisCO will fix carbon dioxide. However, if the carbon dioxide concentration is low, RuBisCO will bind oxygen instead of carbon dioxide. This process, called photorespiration, uses energy, but does not produce sugars.

RuBisCO oxygenase activity is disadvantageous to plants for several reasons:

- One product of oxygenase activity is phosphoglycolate (2 carbon) instead of 3–phosphoglycerate (3 carbon). Phosphoglycolate cannot be metabolized by the Calvin–Benson cycle and represents carbon lost from the cycle. A high oxygenase activity, therefore, drains the sugars that are required to recycle ribulose 5–bisphosphate and for the continuation of the Calvin–Benson cycle.
- Phosphoglycolate is quickly metabolized to glycolate that is toxic to a plant at a high concentration; it inhibits photosynthesis.

- Salvaging glycolate is an energetically expensive process that uses the glycolate pathway, and only 75 per cent of the carbon is returned to the Calvin–Benson cycle as 3–phosphoglycerate. The reactions also produce ammonia (NH_3), which is able to diffuse out of the plant, leading to a loss of nitrogen.A highly simplified summary is:

2 glycolate + ATP → 3–phosphoglycerate + carbon dioxide + ADP +NH_3

The salvaging pathway for the products of RuBisCO oxygenase activity is more commonly known as photorespiration, since it is characterized by light–dependent oxygen consumption and the release of carbon dioxide.

PHOTOSYNTHETIC RESPONSES TO TEMPERATURE

When photosynthetic rate is plotted as a function of temperature, the curve has a characteristic bell shape. The ascending arm of the curve represents a temperature- dependent stimulation of photosynthesis up to an optimum; the descending arm is associated with deleterious effects, some of which are reversible while others are not. Temperature affects all biochemical reactions of photosynthesis, so it is not surprising that the responses to temperature are complex. We can gain insight into the underlying mechanisms by comparing photosynthetic rates in air at normal and at high CO_2 concentrations.

At high CO_2, there is an ample supply of CO_2 at the carboxylation sites, and the rate of photosynthesis is limited primarily by biochemical reactions connected with electron transport. In these conditions, temperature changes have large effects on fixation rates. At ambient CO_2 concentrations, photosynthesis is limited by the activity of rubisco, and the response reflects two conflicting processes: an increase in carboxylation rate with temperature and a decrease in the affinity of rubisco for CO_2 as the temperature rises. These opposing effects dampen the temperature response of photosynthesis at ambient CO_2 concentrations.

Respiration rates also increase as a function of temperature, and the interaction between photorespiration and photosynthesis becomes apparent in temperature responses. Changes in quantum yield as a function of temperature in a C_3 plant and in a C_4 plant. In the C_4 plant the quantum yield remains constant with temperature, reflecting typical low rates of photorespiration. In the C_3 plant the quantum yield decreases with temperature, reflecting a stimulation of photorespiration by temperature and an ensuing higher energy demand per net CO_2 fixed.

At low temperatures, photosynthesis is often limited by phosphate availability at the chloroplast. When triose phosphates are exported from the chloroplast to the cytosol, an equimolar amount of inorganic phosphate is taken up via translocators in the chloroplast membrane. If the rate of triose phosphate utilization in the cytosol decreases, phosphate uptake into the chloroplast is inhibited and photosynthesis becomes phosphate limited. Starch synthesis and sucrose synthesis decrease rapidly with temperature, reducing the demand for triose phosphates and causing the phosphate limitation observed at low temperatures.

The highest photosynthetic rates seen in temperature responses represent the so-called *optimal temperature response*. When these temperatures are exceeded, photosynthetic rates decrease again. It has been argued that this optimal temperature is the point at which the capacities of the various steps of photosynthesis are optimally balanced, with some of the steps becoming limiting as the temperature decreases or increases. Optimal temperatures have strong genetic and physiological components.

Plants of different species growing in habitats with different temperatures have different optimal temperatures for photosynthesis, and plants of the same species, grown at different temperatures and then tested for their photosynthetic responses, show temperature optima that correlate with the temperature at which they were grown. Plants growing at low temperatures maintain higher photosynthetic rates at low temperatures than plants grown at high temperatures.

These changes in photosynthetic properties in response to temperature play an important role in plant adaptations to different environments. Plants are remarkably plastic in their adaptations to temperature. In the lower temperature range, plants growing in alpine areas are capable of net CO_2 uptake at temperatures close to 0°C; at the other extreme, plants living in Death Valley, California, have optimal rates of photosynthesis at temperatures approaching 50°C.

Photosynthetic activity in the intact leaf is an integral process that depends on many biochemical reactions. Different environmental factors can limit photosynthetic rates. Leaf anatomy is highly specialized for light absorption, and the properties of palisade and mesophyll cells ensure uniform light absorption throughout the leaf. In addition to the anatomic features of the leaf, chloroplast movements within cells and solar tracking by the leaf blade help maximize light absorption. Light transmitted through upper leaves is absorbed by leaves growing beneath them. Many properties of the photosynthetic apparatus change as a function of the available light, including the light compensation point, which is higher in sun leaves than in shade leaves.

The linear portion of the lightresponse curve for photosynthesis provides a measure of the quantum yield of photosynthesis in the intact leaf. In temperate areas, quantum yields of C_3 plants are generally higher than those of C_4 plants. Sunlight imposes a substantial heat load on the leaf, which is dissipated back into the air by long-wavelength radiation, by sensible heat loss, or by evaporative heat loss. Increasing CO_2 concentrations in the atmosphere are increasing the heat load on the biosphere. This process could cause damaging changes in the world's climate, but it could also reduce the CO_2 limitations on photosynthesis. At high photon flux, photosynthesis in most plants is CO_2 limited, but the limitation is substantially lower in C_4 and CAM plants because of their CO_2-concentrating mechanisms. Diffusion of CO_2 into the leaf is constrained by a series of different points of resistance.

The largest resistance is usually that imposed by the stomata, so modulation of stomatal apertures provides the plant with an effective means of controlling water loss and CO_2 uptake. Both stomatal and non-stomatal factors affect CO_2 limitations on photosynthesis. Temperature responses of photosynthesis reflect the temperature sensitivity of the biochemical reactions of photosynthesis and are most pronounced at high CO_2 concentrations.

Because of the role of photorespiration, the quantum yield is strongly dependent on temperature in C_3 plants but is nearly independent of temperature in C_4 plants. Leaves growing in cold climates can maintain higher photosynthetic rates at low temperatures than leaves growing in warmer climates. Leaves grown at high temperatures perform better at high temperatures than leaves grown at low temperatures do. Functional changes in the photosynthetic apparatus in response to prevailing temperatures in their environment have an important effect on the capacity of plants to live in diverse habitats.

9

Waged Agricultural Workers

LABOUR IN THE CROP FIELDS

Waged agricultural workers are the women and men who labour in the crop fields, orchards, glasshouses, livestock units, and primary processing facilities to produce the world's food and fibres. They are employed on small- and medium-sized farms as well as large industrialized farms and plantations. They are waged workers because they do not own or rent the land on which they work nor the tools and equipment they use and so are a group distinct from farmers. Such workers do not form a homogeneous group. Their terms and conditions of employment vary tremendously, creating diverse categories: permanent (full-time) agricultural workers; temporary or casual agricultural workers; seasonal agricultural workers; migrant agricultural workers, piece-rate workers; or workers receiving some form of 'in-kind' payment. There are also many indigenous agricultural workers who are part of the employed workforce. Agricultural workers work for some kind of 'wage', whether cash payment, in kind payment, or a combination of these. They work within an employment relationship, be it with a farmer, farming or plantation company, or labour contractor or sub-contractor.

The demand for agricultural labour fluctuates with the seasons and this is reflected in the nature of the workforce. Hours of work tend to be extremely long during planting and harvesting, with shorter hours at off-peak times. During rush periods, field work can go from dawn to dusk, with transport time to and from the fields in addition.

The intensity of the work offers little chance for rest breaks; the length of the working day offers insufficient time for recuperation. Payment systems can exacerbate this situation. As minimum wages tend to be low or non-existent for agricultural workers in general and many casual, temporary or seasonal workers are paid at least in part on a piece work basis – *i.e.* per kilo of crop picked, row weeded, or hectare sprayed, there is a strong financial incentive for them to extend their working time to the maximum so as to enhance their earnings.

Much agricultural work is by its nature physically demanding, involving long periods of standing, stooping, bending, and carrying out repetitive movements in awkward body positions. The risk of accidents is increased by fatigue, poorly designed tools, difficult terrain, exposure to the elements and poor general health. Even when technological change has brought about a reduction in the physical drudgery of agricultural work, it has introduced new risks, notably associated with the use of sophisticated machinery and the intensive use of chemicals often without appropriate safety measures, information and training. Unsurprisingly, the level of fatal and serious accidents and illness is high. Yet, agricultural workers are among the least well protected in terms of access to health care, workers' compensation, longterm disability insurance and survivors' benefits.

PERMANENT AGRICULTURAL WORKERS

Full-time, permanent agricultural workers receive more job security, relatively higher wages, better housing and better health and work benefits than do other waged agricultural workers. However, this does not mean that such workers are well-paid, relative to the average wage levels in a given country. Wages in rural areas, both in cash terms and in real terms, are generally lower than in cities, and the hours of work are longer. Many full-time workers in agriculture live on or below the poverty line. Permanent contracts are the least common form of contract and their share in total agricultural employment has been declining in most countries. There is a trend away from full-time employment to more casual and seasonal employment, often referred to as the "casualization" or "flexibilization" of employment, with little or no social protection.

MAJORITY OF WAGED AGRICULTURAL WORKERS

The majority of waged agricultural workers in most developing countries and in some developed countries are employed on a seasonal and often a casual or temporary basis. Casual work refers to those employed and paid at the end of each day worked or on a task basis. Temporary work refers to those employed for a specific but limited period of time. Most seasonal, casual or temporary workers do not receive any form of social security or unemployment benefit, holidays with pay, or sickness or maternity leave. Indeed, many full-time waged agricultural earners lack these same benefits. An increasing number of women workers are employed as casual or temporary workers. Furthermore, jobs are often classed as casual or temporary even if there is in reality continuous employment. The practice of rotating individual workers so as to deny them the benefit of permanent employment status is also quite prevalent. In the cut flower industry, for example, data provided by the Ugandan National Union of

Plantation and Agricultural Workers affirms that the majority of casual workers are women.

Over 70 per cent of this workforce is casually employed without job security and other benefits such as annual leave. The women workers are mostly employed in harvesting and in the grading sections. A fulltime worker is paid a total package of 70,000 Uganda shillings ($35 US) per month while a casual worker earns 1,500 Uganda shillings per day (75 cents US). The trend towards casual and temporary labour is encouraged by, amongst other factors, unpredictable weather conditions, unstable market demand for produce, and labour laws which require that certain benefits, such as notice pay, leave allowances, and medical attention, be provided to seasonal and permanent employees.

To counter this development towards casual and temporary labour, unions in the plantation sector in many countries such as, for example, Kenya, have worked to include provisions in collective bargaining agreements limiting the length of time during which employees may be engaged on a seasonal, casual or temporary basis.

TYPES OF EMPLOYMENT RELATIONSHIPS

Migrant workers are found in all types of employment relationships as casual, temporary, seasonal or even full-time workers. They may be migrant workers from a different part of a country, or foreign workers. Wherever they come from, migrant workers are always heavily disadvantaged in terms of pay, social protection, housing and medical protection. Globalization and structural adjustment have transformed agriculture in many countries into an export-oriented sector heavily dependent on migrant labour.

According to a trade union consultation on Irregular Migration and Human Trafficking in Europe, 2003, it is essentially the power of the large supermarket chains that forces farmers to produce at very low cost. Farmers respond to the pressure by decreasing their labour costs, thus passing the burden on to the workers. Low pay and status for hard work is not attractive for the national workforce. Therefore farmers rely increasingly on migrant labour.

In some areas, as in the Almeria region of Spain, more than 90 per cent of workers in agriculture are immigrants. Labour migration is one of the major consequences of growing work flexibility, casual employment, low pay, bad working conditions and poverty. This labour mobility is very significant throughout the world. The migrant labour force often consists in reality of whole families, although formally only the head of the family is employed. In many countries, children of migrant and seasonal workers work next to their parents but do not figure on the payroll. As much work is paid on a "piece-rate" basis, migrant and seasonal workers need their children to work in order to achieve a living wage.

AGRICULTURAL WORKFORCE

Indigenous agricultural workers form part of the agricultural workforce in many countries. Having been forced off their own lands, they often end up as agricultural workers, usually under poor employment and working conditions. In many countries in Latin America, for example, the living and working conditions of indigenous workers and their families are below the average for workers belonging to what is referred to as the "predominant society", including non-indigenous rural workers. These differences are evident in access to education, social security, levels of trade union membership, salaries and occupational health and safety.

DEFINITIONS OF CATEGORIES OF WORKERS

These categories of waged workers often overlap. The lack of standardization of definitions impacts on accurate calculation of the number of workers. For example, farmers in the Western Cape, South Africa, employ women workers differentiated between permanent, seasonal and temporary or casual workers. However, definitions of permanent workers differed greatly and many farmers defined casual workers as seasonal and vice versa. There was in fact little consistency in terms from farm to farm, or within a sector or region.

WAGE-DEPENDENT SMALL FARMERS

Many small farmers regularly work on a farm or plantation for part of the year to supplement their meagre incomes. Their annual income depends on waged work as a regular source of revenue. In reality, they are part farmer/ part waged worker. A 1998 study of the Mexican agricultural labour market, for example, estimated that as many as 4.8 million farmers, forming 78.3 per cent of the rural labour force, were engaged as waged workers for at least part of the year.

This phenomenon of wage-dependent smallholders, though well-known, rarely forms part of agricultural programmes to empower small farmers. The need to increase smallholders' wages and improve working conditions is rarely taken into account in poverty eradication programmes and strategies.

Waged Agricultural Workers and Small Farmers

There are many common issues in sustainable agriculture and rural development, and food security, affecting waged agricultural workers and small farmers and many similarities in their needs and roles, especially where farmers are regularly wage-dependent. Similar interests include, for example, the right of freedom of association and organization and improving health, safety and environmental conditions on farms and plantations. As rural employment becomes increasingly casualized, the distinction between waged agricultural workers and the rest of the rural working population becomes less and less clear-cut.

As a result, waged agricultural workers have become potential allies of disadvantaged rural groups such as small, subsistence farmers, tenants and sharecroppers, the unemployed and the landless. They are engaged in similar activities, share the same environment and often come from the same household. Although there are sometimes tensions and conflicts between these different actors, relations based on solidarity and mutual support can be developed as their interests converge and common strategies are elaborated to tackle common problems. The potential links can be illustrated by the case of a sugar plantation and processing facility in Kenya managed by a multinational enterprise (MNE). This sugar company employs 3,200 permanent employees on its own nucleus plantation, but contracts, buys in and processes sugar from a network of 65,000 outgrowers, *i.e.* small farmers, grouped under their own company. The two entities – the sugar company and the outgrower network – are totally interdependent.

WORLDWIDE PHENOMENON AND CHILD LABOURERS IN AGRICULTURE

Child labour is a worldwide phenomenon. An estimated 246 million children around the world carry out work that harms their well-being and hinders their education, development and future livelihoods. Child labour is work which, by its nature and/or the way it is carried out, harms, abuses and exploits the child or deprives the child of an education.

Child labour takes many different forms, and whilst the long-term goal is the elimination of all forms of such labour, a priority is to eliminate without delay the *worst forms of child labour* as defined by the ILO Worst Forms of Child Labour Convention, 1999 (No.182). In this Convention, the term "child" applies to all persons under the age of 18. Seventy per cent of all child labourers work in agriculture.

While many children have traditionally been employed in family enterprises, children also work in large-scale commercial plantations and in agriculture as migrant farm workers. They usually work alongside their parents, often in situations of indirect employment where only the head of the family is actually employed, but where he or she is paid according to the amount of fruit or vegetables that are turned in at the end of the day.

With more hands to help, the family makes more money. Participating in home-based agriculture helping the family is a completely different situation from working during harvesting on a piece-rate basis, as a member of a migrant worker's family in a foreign country. These migrant children work long hours, sometimes over 60 hours a week. In the Philippines, for example, child labourers work 10 hours per day from Monday through Saturday with only short breaks and half a day on Sunday.

The children earn less than one dollar daily. They weed, cultivate, turn soil, fix canals, harvest and apply pesticides. Children weed, cut cane and apply fertilizers on sugar plantations. Sometimes, they are as young as 7-8 years old and begin to cut the cane at age 12. Children are injured by using sharp knives, and are poisoned by the use of dangerous fertilizers.

An estimated 171 million children aged 5-17 – in all economic sectors - work in dangerous, hazardous conditions that could result in them being killed or injured (often permanently) and/or made ill (often permanently). Work which results in children being killed, injured or made ill as a consequence of poor safety and health standards and working conditions/arrangements is called "hazardous child labour".

Child labourers in agriculture are especially at risk, as agriculture is one of the three most hazardous sectors in which to work, along with construction and mining. Children's work is often invisible, because they assist their parents in task or piecework.

Because this work is not recognized, nor easily recorded in statistics, it is goes largely unnoticed, and is not taken into account in policy.

An estimated 22,000 children are killed every year at work in all sectors. No figures for child accidents or ill-health due to work for any sector are currently available.

But every year there are 270 million work accidents and 160 million cases of ill-health due to work, and child labourers figure amongst these statistics. The risks arising from hazards in the workplace are much greater for children. They are more vulnerable due to a variety of factors including:

- Lack of work experience.
- Lack of knowledge about hazards and risks, and on risk prevention and control measures.
- Children's bodies are still growing and their minds developing. Frequent awkward or heavy lifting and repetitive strains, for example, can permanently injure growing spines or limbs, especially if poorly designed equipment is being used. Skin, eye, respiratory or nervous problems occur in children exposed to pesticides, and children are vulnerable to much lower levels of exposure than adults. There may well be chronic long-term health effects from exposure to pesticides that will not show up until the child is an adult.
- Children may also experience conflict because they are expected to work like adults, whilst still behaving obediently. The aim is the elimination of hazardous child labour, with no child under 18 undertaking hazardous work, with very limited exemptions for young workers, aged 16-17 – for training purposes.

Prevalence of Child Labour in Agriculture

The prevalence of child labour in agriculture also undermines decent work, sustainable agriculture and food security as it maintains a cycle where household income is insufficient to meet the needs of families. Children work as cheap labour because their parents do not earn enough to support the family or to send their children to school.

Working children represent a plentiful source of cheap labour. Child labour undermines the bargaining power of agricultural workers and their trade unions and weakens the possibility of negotiating a fair and decent wage for adult workers. Measures have to be taken to improve workers' wages and farmers' incomes so that their children are not obliged to work to try and help get their families enough to live on.

Improving conditions for waged agricultural workers will in the long term reduce child labour by ensuring that workers have the economic means to support their families. The overall aim is to eliminate child labour and to provide quality education for children and more and better jobs for their parents. Waged agricultural workers, farmers and their organizations have vital roles to play in helping to realize these aims.

YOUNG WORKERS

Article 16 of the ILO Safety and Health in Agriculture Convention, 2001 (No.184) states that "the minimum age for assignment to work in agriculture which by its nature or the circumstances in which it is carried is likely to harm the safety and health of young persons shall not be less than 18 years". However, national laws or regulations or the competent authority may, after consultation with the representative organizations of employers and workers concerned, authorize the performance of work as from 16 years of age on condition that appropriate prior training is given and the safety and health of the young workers are fully protected. The "employment relationship" involved in the different forms of waged employment is often complex and difficult to understand, especially where casual labour is employed or labour is supplied through labour contractors.

The employment relationship is a notion which creates a legal link between a person called the "employee" and another person called the "employer" to whom she or he provides labour or services under certain conditions in return for remuneration. The concept of an employment relationship is common to all legal systems and traditions but the obligations, rights and entitlements associated with it vary from country to country. One of the consequences associated with changes in the structure of the labour market, the organization of work and the deficient application of the law is the growing phenomenon of workers who are in fact employees but find themselves without the protection of an employment relationship. This form of false self-employment is more common in less formalized economies. However, many countries with well-structured labour markets are also experiencing an increase in this phenomenon. Some of these developments are new; some have existed for many decades.

STRUCTURE OF THE LABOUR MARKET

Changes in the structure of the labour market and in the organization of work are leading to changing patterns of work both within and outside the framework of the employment relationship. In some situations, it may be unclear whether the worker is an employee or genuinely self-employed. Disguised employment occurs when the employer treats a person who is an employee as other than an employee so as to hide her or his true legal status. False self-employment, false subcontracting, the establishment of pseudocooperatives, false provision of services and false company restructuring are amongst the most frequent means that are used to disguise the employment relationship.

The effect of such practices can be to deny labour protection to the worker and to avoid costs that may include taxes and social security contributions. An ambiguous employment relationship exists whenever work is performed or services are provided under conditions that give rise to an actual and genuine doubt about the existence of an employment relationship. In an increasing number of cases, it is very difficult to distinguish between dependent and independent work, even where there is no intent to disguise the employment relationship. In this respect, it is acknowledged that in many areas the distinction between employees and independent workers has become blurred. One of the characteristics of some new forms of work is the autonomy or greater independence of employees.

Clarity and predictability in the law are in the interests of all concerned. Workers and employers should know their status and, consequently their respective rights and obligations under the law. It is in the interest of all the labour market actors to ensure that the wide variety of arrangements under which work is performed or services are provided by a worker can be put within an appropriate legal framework. Clear rules are indispensable for fair governance of the labour market.

WAGED AGRICULTURAL WORKERS

Waged agricultural workers are a very large occupational group.

- There are an estimated 450 million waged agricultural workers out of a total workforce in agriculture of some 1.1 billion. They account in fact for over 40 per cent of the total agricultural labour force and play a major role in feeding the world and in fibre production.
- Waged employment is now a central feature of employment in rural areas, and the number of waged workers is increasing in most regions of the world.

- Women waged agricultural workers account generally for 20-30 per cent of the waged workforce, rising to 40 per cent in Latin America and the Caribbean. The numbers of women in waged employment are also increasing in most regions. New jobs are usually in export-orientated agriculture like cut flowers and vegetable growing and packing. These jobs are often classed as temporary or seasonal even if there is, in reality, continuous employment.
- Waged workers are often poorly paid, with wages well below those earned by industrial workers. They and their families often live below the poverty line, and they form part of the core rural poor in many parts of the world. Millions of these workers earn the lowest wages in the rural sector, lower even than the amount required to subsist.
- Their employment is often unstable and temporary. Employment problems have increased as the impact of globalization has led to less and less permanent labour and a more casualized and marginalized workforce (often hired through labour contractors or subcontractors). Migrant workers face particular difficulties.
- Working and living conditions are often poor.
- Agricultural workers earn their living in an industry ranked as one of the three most hazardous along with mining and construction. They face a wide variety of hazards at work ranging from, for example, dangerous machinery, unsafe electrical wiring and appliances, livestock-transmitted diseases and falls from heights, to exposure to toxic pesticides. Every year, large numbers of waged agricultural workers and farmers are killed (170,000 in 1997 according to ILO estimates), injured or made ill as a result of their work.
- In many parts of the world, agricultural workers are denied fundamental human rights: the rights to freedom of association, to organize and to collectively bargain with employers. Rural workers are more subject to forced labour than other categories of workers. Yet decent conditions of work and respect for the fundamental rights of agricultural workers are essential to sustainable development.
- Agricultural workers in many countries are typically excluded from active participation in decision-making processes with employers and government.
- HIV/AIDS is also devastating the agricultural labour force in many parts of the world. HIV/AIDS is a workplace issue. The disease not only affects the workforce, but also results in loss of agricultural productive capacity, undermining farms, plantations and the national economy.
- The prevalence of child labour in agriculture undermines sustainable agriculture and food security as it maintains a cycle where household

income is insufficient to meet the needs of families. Most children work because their parents do not earn enough to support the family and to send them to school.

WORKERS AND THEIR TRADE UNIONS IN AGRICULTURE

A trade (or labour) union is an organization created and run by workers to protect and promote their livelihoods and labour rights in workplaces. A free trade union is independent of patronage. It is not a workers' organization established openly or secretly by an employer to control the actions or demands of the workers or an organization promoted by the elite or the authorities for the same purposes. Through collective bargaining with employers, trade unions work to improve their members' remuneration and conditions of work, and to ensure their security of employment. Unions also lobby and work with governments to maintain and improve labour rights, standards and policies.

Workers' membership dues provide the financial basis of unions. Union policies are democratically decided at conferences and congresses by delegates elected by the members. Unions also provide workers with a means of expressing their views on wider societal, economic, ethical and political issues.

The trade unions, federations and confederations found in agriculture include:

National Agricultural Trade Unions

- National agricultural trade unions having a membership composed solely of agricultural workers on commercial farms and plantations. This type of union is normally found in developing countries where agriculture is still the dominant economic activity. Such unions may organize workers across all agricultural sectors, or only have membership in specific crop sectors, *e.g.* tea, sugar, tobacco. These unions are often the largest workers' organization in a country. Plantation agriculture is often synonymous with large unions. Examples include the National Union of Plantation and Agricultural Workers in Uganda, and the Kenya Plantation and Agricultural Workers Union in Kenya.
- National general workers' unions bring together members from different industrial and commercial sectors and are found in countries where the size of the agricultural workforce is relatively small. Waged agricultural workers form sections or trade groups within these large general unions. Such unions are often found in developed countries. Examples include the agricultural workers' sections of Kommunal in Sweden, SiD in Denmark and the Transport and General Workers in the UK.

- There are also agricultural trade unions which are hybrid organizations representing both agricultural wage earners and self-employed farmers. For example, CONTAG in Brazil has some 9 million members (3 million waged workers and 6 million smallholders). Such unions usually have separate organizational and political structures for dealing with the dual types of membership. Sometimes the relationship between the two types of membership in the same organization is not easy, especially where small farmer members also employ waged labour.
- National trade unions – whether agriculture-specific, general or hybrid - usually affiliate to, or in some cases are part of, a national trade union centre. A national centre represents unions organizing workers across different industrial and commercial sectors. There may be more than one national centre in a country.
- Some agricultural unions are organized on a federal state basis such as, for example, in India.
- In some countries, labour laws confine unions to organizing in one company. An example of this is the banana union SITRABI in Guatemala.

International and Regional Levels

National unions may affiliate on a sectoral basis to global union (labour) federations such as the International Union of Food, Agricultural, Hotel, Restaurant, Catering, Tobacco and Allied Workers' Associations (IUF). The IUF network of 336 national affiliated unions organizes workers on farms and plantations, in food and drink manufacturing companies, in hotels, restaurants, and tourism and catering services around the world. The IUF has targeted six major crops – bananas, cocoa, coffee, (cut) flowers, sugar and tea – for developing links between unions along the food chain and bringing together workers in all stages of the production process.

Global union federations work in turn with the International Confederation of Free Trade Unions (ICFTU), which is made up of national trade union centres. The ICFTU has 233 affiliated organizations in 154 countries and territories on all five continents, with a membership of 148 million workers. It has three major regional organizations, APRO for Asia and the Pacific, AFRO for Africa, and ORIT for the Americas. It also maintains close links with the European Trade Union Confederation (ETUC) and Global Union Federations.

Two other international trade union organizations made up of national centres also exist. The World Confederation of Labour (WCL) is an international trade union confederation uniting 144 autonomous and democratic trade unions from 116 countries with over 26 million members. Its head office is located in Brussels, Belgium. Merger talks are now underway between the ICFTU and

WCL to create a new international trade union organization. The World Federation of Trade Unions (WFTU) is located in Prague.

Union Organization in Transnational Corporations in Agriculture

Global union federations such as the IUF represent and organize workers in large multinational enterprises (also referred to as transnational corporations, TNCs for short) in agriculture and along the food chain in food and drink processing companies, catering services, restaurants and hotels/tourist facilities, and in retail and wholesale distribution. National and regional trade unions through their international federations can play important roles in working with such multinational enterprises (MNEs) and monitoring their performance and progress, especially in respect of labour standards, on moving towards sustainable agriculture.

TRADE UNION ORGANIZATION IN AGRICULTURE

The level of trade union representation among agricultural workers, and particularly workers who are not permanent, is generally low in most countries, and particularly among women agricultural labourers. Data on agricultural trade union membership are far from comprehensive. There are many reasons why agricultural workers remain poorly organized. Some of these difficulties are practical or financial, *e.g.* difficulties of organizing over large geographical areas, lack of transport for organizers, low membership dues resulting in only basic union services, and so on.

Often however legal and administrative barriers are placed in the way of workers wishing to exercise their basic human right to freedom of association by becoming unionized and joining the union of their choice. In Turkey, for example, the labour laws were only modernized in 2003 to cover agricultural workers. However, each new union membership application has to be approved by the public notary, for which a substantial fee is charged. This requirement makes it costly for unions, especially poor agricultural ones, to recruit members, and is an obstacle to freedom of association. The situation is even more critical for indigenous workers who are often denied the legal capacity to set up or join such organizations.

The fundamental right of both employers and workers to establish and join organizations of their own choosing as a means of defending their interests and improving their conditions is enshrined in the principle of freedom of association as expressed in the ILO Right of Association (Agriculture) Convention, 1921 (No.11) and the Freedom of Association and Protection of the Right to Organize Convention, 1948 (No.87). Both these Conventions are part of the ILO International Declaration on Fundamental Principles and Rights at Work. The ILO Rural Workers' Organizations Convention, 1975 (No.141) also seeks to extend the principle of freedom of association to those working in agriculture.

However, violations of freedom of association among those working in agriculture are frequent, as evidenced by cases examined by the ILO Committee on Freedom of Association.

Such violations range from regulatory restrictions and practical impediments to physical assaults and the assassination of trade union leaders and officials in rural areas. Cases of forced and bonded labour involving agricultural workers in a number of countries in several regions have been examined regularly by the Committee of Experts. The ICFTU also publishes an annual survey of violations of trade union rights around the world. Furthermore, waged workers engaged by labour contractors or subcontractors often face difficulties in registering as union members. This is causing growing concern, due to the increasing numbers of workers hired by these laboursupplying intermediaries.

DETERMINING EMPLOYMENT AND INCOME LEVELS IN RURAL AREAS

Labour markets play a key role in determining employment and income levels in rural areas. While agriculture cannot be expected to absorb all of the rural labour force, its direct contribution to the generation of employment, including wage employment, and its indirect contribution through greater diversification of the economy, are critical.

Access to labour markets is particularly important for many of the rural poor, as hiring out their labour power may be their sole source of income. Often, the only asset possessed by waged agricultural workers is their labour. Hence the importance of improving the functioning of rural labour markets, as this is the only effective way of improving the productivity of their main asset, and therefore livelihoods of the rural poor.

WAGES AND POVERTY

The three major trends in agricultural labour markets in the 1980s and early 1990s, reflected in data collected by the International Labour Organization (ILO), have been:

- An increase in the share of agricultural waged employment in total rural economic activity;
- An increase in the share of women in agricultural waged employment;
- Growing casualization of agricultural waged labour.

There are more workers in waged employment in agriculture today than at any time. The share of waged employment in agriculture, including the number of wage-dependent smallholders in agriculture, is continuing to increase in virtually all regions, and it is now a central feature of employment and income in rural areas. In other words, as the agricultural sector is undergoing a process of concentration of ownership, leading to fewer and bigger farms with a higher number of waged workers - as opposed to a sector based on smaller production units and self-employed farmers - waged workers become central to agricultural production. In 1996, the highest average shares of waged employment were found in Eastern and Central Europe and Central Asia (over 80 per cent), followed by Latin America and the Caribbean (over 50 per cent), Asia (over 40 per cent) and sub-Saharan Africa (30 per cent), Organisation for Economic Cooperation and Development (OECD) countries (35 per cent) and Near East and North Africa (25 per cent).37 Also, the share of wage employment seems to move inversely with the share of the agricultural labour force - that is, the smaller the labour force in agriculture, the higher the share of waged employment. As the economy absorbs more labour in other sectors, and the agricultural labour force shrinks, production units in agriculture are transformed from small family holdings to large units relying on wage labour; in other words, the number of smallholders decreases.

WOMEN AND WAGED EMPLOYMENT

Significantly, the share of women waged agricultural workers has also been rising in all regions and women now account for 20-30 per cent of total agricultural wage employment. In Latin America and the Caribbean, the figure is 40 per cent.39 In African countries, this percentage is likely to be higher; however reliable data are difficult to obtain given the prevalence of informal labour arrangements.

INDIAN TRADE UNIONS OF THE IMPACTS OF GLOBALIZATION ON AGRICULTURAL WORKERS

A study by Indian trade unions of the impacts of globalization on agricultural workers confirms the increasing feminisation of the workforce. Male workers have shown a greater tendency to migrate to non-agricultural work. Women have replaced them. The unions observe that landowners also seem to prefer women workers, as they are able to pay them less and find them a docile and dependent workforce. Thus, wage discrepancies between men and women have continued. Increasing employment opportunities and higher levels of employment and earnings for women (not uniquely in agriculture) are critical for empowerment and food security.

A recent study in the Latin American and Caribbean region demonstrates that increasing the share of women's income in the household considerably improves family and social welfare, given women's likelihood to invest more than men would in children's potential as human capital. Increasing women's earnings and share of family income has also been shown to empower women by strengthening their bargaining power in the household. Women's earnings are especially important, given the growing number of households headed by females.

WORLDWIDE TREND TOWARDS CASUALIZATION OF WORK

The worldwide trend towards casualization of work, brought on in large measure by deregulation, globalization and competitive pressure to reduce labour and production costs, is changing the nature of agricultural employment in many parts of the world. Global trade and economic pressures are eroding the already low levels of protection of agricultural workers in terms of wage levels, employment security, health, safety and environmental standards and social protection.

Casualization means there is a trend away from permanent employment contracts and workers are increasingly being employed as temporary or casual labour on short-term, daily or seasonal contracts with poorer pay and working conditions. There is also a developing trend by agricultural employers towards outsourcing of work - that is, the process of relocating outside the enterprise the production processes previously carried out in-house.

The most serious problem is that of labour hired through or by contractors. This mainly applies to casual workers, especially migrants. Employers, who need to be able to count upon a supply of labour in periods of peak demand, increasingly call on the services of labour contractors specialized in the recruitment, transport and management of waged agricultural workers. This process involving middlemen undermines the whole employment relationship, creating a grey area around the employer's responsibilities and leading to a disregard for labour legislation.

Conditions of employment are generally insecure and the labour contractors frequently abuse their authority over the workers by asking for commissions, over-charging for transport, housing and food, holding back wages and imposing debt slavery. Most agricultural trade unions in India see the present global restructuring of the economy as a continuation of the commercialization of agriculture that has been taking place since the 1960's. They see the most significant impact of globalization on agricultural workers as their displacement from work and livelihood. This has happened due to the adoption of labour-displacing mechanization and other new technologies and changes in crop patterns, with less labour-intensive crops.

UNEMPLOYMENT AND UNDEREMPLOYMENT IN RURAL AREAS

Unemployment and underemployment in rural areas are also major causes of poverty among waged agricultural workers and are regular and significant features of their lives. If full employment is assumed to be 260 days per year (52 five-day weeks, excluding leave and holidays), waged agricultural workers typically find employment for 175 days and are available for work (*i.e.* unemployed) for 85, with little income to sustain them between seasons. These large pools of unemployed or underemployed labour keep down wages locally.

Another feature of rural labour markets is the high incidence of hidden unemployment, which is the underestimation of unemployment levels in labour statistics, where only those who are "actively looking for work" are counted as unemployed. As rural labour markets are generally not functioning smoothly, information is imperfect and transaction costs are high. There is a tendency for those who are not in employment to give up looking for jobs, and for the employed to work for less time that they would like.

ADOPTION OF SUSTAINABLE AGRICULTURE PRACTICES ON EMPLOYMENT GENERATION AND LABOUR MARKETS

A key question concerns the impact of adoption of sustainable agriculture practices on employment generation and labour markets. More data and further research are needed as there is only very limited evidence of either positive or negative impacts at present.

A research survey of sustainable agricultural practices worldwide by Essex University demonstrates that such practices can have a significant impact on labour markets. Some sustainable agricultural practices resulted in increased on-farm demand for labour (*e.g.* water harvesting in Niger); others actually reduced labour demand (*e.g.* zero-tillage in Brazil); some resulted in the opening up of whole new seasons for agricultural production, particularly in dryland contexts, through improved harvesting of rainfall, leading to much greater demand for labour.

Reverse migration, from urban to rural areas, could also occur when waged labour opportunities increased through sustainable practices (*e.g.* watershed improvements), when more productive agriculture led to higher wages and employment, when there were higher returns to agriculture, and when there were overall improvements in village conditions, such as infrastructure and services.

The real issue therefore is whether sustainable agriculture practices can be used to redress imbalances in labour supply and demand, and whether they can raise labour productivity.

It is important to stress that, given the marked gender division of labour in some agricultural tasks, an evaluation of sustainable practices' impact on labour requires an analysis of the differential impact on men and women.

This is because a technology determining savings or increased used of labour could have a different impact on men's and women's workload, depending on who will end up carrying out the tasks that are reduced/increased/substituted with the adoption of a new sustainable agriculture practice.

Poverty line of Wages

The work that waged agricultural workers do is often badly paid and as a result many live below the poverty line. Also, there is a large gap between the average earnings of industrial workers and their agricultural counterparts. For example, data from Kenya showed that in 2000, agriculture and forestry offered wage employment to 311,000 workers out of a total regular waged workforce of 1,676,800 nationwide.

Agriculture and forestry contributed more than 8.1 per cent of the total wage earnings in 2000. However, wage earnings per employee in private sector agriculture and forestry were the lowest of any sector. At Kenya shillings (Ksh) 66,000 per annum, average wage earnings for employees in agriculture were less than 40 per cent of average wage earnings among private sector employees as a whole.

The income determinants for waged workers are:

- For full-time workers: wage level, number of wage earners in household;
- For casual workers: number of days worked, wage level, number of wage earners in household; share of income derived from plantation/ farm wage work.

Income gaps between full-time wage earners and casual workers can be quite wide, not only because the number of days worked is lower, but also because benefits such as housing, education and health care are not provided.

MAJOR ELEMENTS AFFECTING WAGE LEVELS IN AGRICULTURE

Five major elements affecting wage levels in agriculture can be identified:

Food Prices and Food Security

Worker households often spend over 70 per cent of their cash wage on food. Rising food prices can push significant numbers of waged workers and their families below, or even further below, the poverty line. Improving earning power and livelihoods and ensuring food security are closely linked issues for agricultural workers and their trade unions.

Whilst noting that targets have been set at successive World Food Summits to improve food security globally, Indian agricultural trade unions see "food security for all" as a distant dream, in spite of huge stocks of food grains in their country.

Because of low earnings, agricultural workers lack the purchasing power to buy sufficient food from the market. Furthermore, changes in crop patterns from staple food grains towards higher priced cash crops have worsened food insecurity for many worker households. The unions believe that dependence on imported food is likely to increase, eventually leading to lack of food sovereignty.

Labour Supply

Real increases in wages are closely associated with rising labour absorption in agriculture, or land-augmenting agricultural practices based on higher cropping intensity, stimulated by investment in irrigation, such as happened in Bangladesh and parts of India in the 1980s. Robust agricultural growth, fuelled by exports, can also be an important means of real wage increases.

Non-Farm Employment

Rural non-farm employment in the form of emergency relief or povertytargeted labour-intensive public works has been promoted by many governments over the years. The ILO has long advocated diversification of rural employment through such labour-intensive works programmes in view of the benefits accruing to the poor, both directly through employment and income generated by the public works programmes and indirectly, through the employment effects induced by the infrastructure created and the increased demand for locally-produced goods. In addition, producers of simple consumer goods may also be located in rural areas, particularly those making use of local raw materials. The development of agro-ecological tourism has also become important in some areas.

Rural non-farm activities are generally small-scale and labour-intensive, using local materials and generally catering to local demand. Such activities can provide supplementary employment to that section of the rural labour force significantly underemployed during the lean agricultural season. Despite declining levels of labour absorption in agriculture in some countries, aggregate rural employment can rise with the expansion of non-farm activities and result in lower levels of poverty, as has been demonstrated especially in countries such as China, northern India, Pakistan and some Latin American countries.

Agricultural Growth

The classical "trickle-down" argument is that agricultural productivity growth can translate in a sustainable reduction of hunger and poverty, as farm incomes rise following productivity increases. Increased farmers' incomes and higher agricultural workers' wages create increased demand for basic nonfarm products and services in rural areas. These include: tools, blacksmithing, carpentry, clothes, processed food bought from roadside kiosks. These goods and services are often difficult to trade over long distances. They tend to be produced and provided locally, usually with labour-intensive methods, and so have great potential to create employment and alleviate poverty. FAO surveys in four African countries showed that between one-third and two-thirds of income increases in rural areas were spent on local goods and services. However, according to a recent study by Indian trade unions on the impacts of globalization on agricultural workers there has not been any automatic trickledown effect of increased productivity on wages in the agricultural sector. Positive changes in wage rates seem instead to be associated with bargaining by agricultural workers or with the availability of alternative work opportunities. Despite the nominal gains, real wages have in fact deteriorated as consumption goods that were earlier available free of cost (*e.g.* water, fuel, fodder, wild vegetables, and river fish) are now becoming marketable commodities.

Minimum Wages

National or sectoral statutory or negotiated minimum wages aim to protect the most vulnerable and lowest paid workers so as to guarantee a living wage to all workers. The ILO has long been advocating a system of minimum wages to protect the lowest-paid workers from undue exploitation.

As collective bargaining is often weak in agriculture, some form of national wage-fixing machinery, involving government, can be important to set minimum wage levels. Such wage-fixing machinery varies according to national systems and practice but is usually characterised by having a legal framework with some form of third-party arbitration between employers and trade unions, often binding, *e.g.* Wages Boards.

Enforcement of minimum wages is widely thought to be difficult if not impossible in rural areas in view of the extent of surplus labour and widespread unemployment. The largely informal nature of labour contracts in agriculture seems to preclude the possibility of enforcing a non-market determined minimum wage.

Unfortunately, no comprehensive survey or data sources are available to enable any assessment of minimum wages in agriculture.

Unions have resisted attempts by governments at wage deregulation and abolition of wage-fixing machinery. In the United Kingdom (UK), for example, the agricultural workers section of the Transport and General Workers Union has vigorously defended the UK Agricultural Wages Board against government attempts at deregulation and called for a minimum wage.

POVERTY AMONG WAGED WORKERS AND SUSTAINABILITY

Currently, one in four people in developing countries lives in extreme poverty - defined as subsisting on less than US$ 1 a day. Figures published by the FAO in 2002 show that 776 million people in developing countries remain undernourished - about one person in six.

Undernourishment also deepens other aspects of poverty, by reducing the capacity for work and resistance to disease, and by affecting children's mental development and educational achievements.

About 70 per cent of poor people in developing countries live in rural areas and depend directly or indirectly on agriculture for their livelihoods. Waged agricultural workers and their families are part of the core rural poor, though precise recent data are lacking as poverty estimates generally distinguish only between rural and urban areas. High levels of poverty among agricultural wage labourers are a source of concern, as their share in the rural labour force is increasing in virtually all regions.

Poverty is not confined to subsistence farming but is a feature of commercial agriculture as well. High incidences of poverty are common among waged agricultural workers on plantations. In 1987, for example, Indonesian agricultural workers accounted for less than 10 per cent of the rural workforce but recorded the highest incidence of poverty, with 38 per cent of all agricultural waged workers below the poverty line.

Waged agricultural workers are among the occupational groups with the highest incidence of poverty in a number of countries, often over 60 per cent. This can vary by region. Data from the ILO showed that in Africa and Asia, waged workers experienced consistently higher rates of poverty than the rural population in general, whereas in Latin America, with the exception of Chile and Mexico, the reverse seems to be true.

In terms of poverty elimination amongst waged agricultural workers and their families, the critical variables are wage levels and employment. For these reasons, waged agricultural workers along with small farmers should be a key target group of the United Nations Millennium Development Goal 1: Eradicate extreme poverty and hunger:

- Halve, between 1990 and 2015, the proportion of people whose income is less than one dollar a day
- Halve, between 1990 and 2015, the proportion of people who suffer from hunger.

However, it is not clear that waged agricultural workers are seen as a key target group. They are hardly ever mentioned in United Nations documentation outside of the ILO. If they are never mentioned, how can their needs as part of the core rural poor be assessed and programmes to eradicate their poverty and hunger be developed?

RISING SHARE OF WAGED LABOUR IN AGRICULTURE

With a rising share of waged labour in agriculture and an increasing percentage of total output marketed domestically or internationally, concern for better conditions of employment at work should at least match concern for product quality standards.

Improvements in conditions of employment at work are a necessary prerequisite to modernization and the opening of markets and they can stimulate economic efficiency by raising labour productivity.

Such improvements should be considered as an investment rather than a social cost.

Labour Law and International Labour Standards

Compared to those working in other economic sectors, many agricultural workers are only poorly protected by national labour law. In some cases, this is because the scope of the relevant legislation – a Factories Act, for example – is limited to industrial enterprises and their workforce. In some countries, the agriculture sector is specifically excluded from the scope of general labour legislation. Where it exists, protective legislation may not be fully applicable to the agriculture sector, may be out of date or may simply not be applied. Application of general labour laws can be problematic in rural areas, where employers and workers may be less familiar with the details of the law and compliance may be considered impractical, particularly in small-scale farms and enterprises. Even where specific legislation has been enacted which accommodates the special characteristics of agricultural work – for example, with regard to working time arrangements, wage structure, and the provision of housing in remote areas – inspection and enforcement tend to be weak.

Promoting Rights at Work

International labour standards are comprised of ILO Conventions and Recommendations which together form the International Labour Code. ILO standards span most subjects relevant to labour law and the social aspects of development and thus provide guidance to member States for the improvement of national labour legislation and social policy. ILO standards are universal in character and are formulated with sufficient flexibility to take account of variations in conditions and practices in countries at different levels of development.

Many ILO Conventions apply to all workers. These include, but are certainly not limited to, ILO core labour standards on freedom of associationvii, the right to collective bargainingviii, non-discrimination, equal pay for men and women workers, the abolition of forced labour, and the elimination of child labour. These core labour standards are often referred to as human rights at work. Not only are they important in their own right, but they also serve as enabling rights. That is, they create conditions to allow access to other rights. Freedom of association is a prime example of that function. The right of workers and employers to establish and join independent organizations of their own choosing creates the basis on which social dialogue between employers and workers can take place, with a view to regulating terms and conditions of employment through collective agreements. Freedom of association is a fundamental human right which paves the way for improvements in social and labour conditions, for example, through collective bargaining.

Despite nearly universal recognition of the right to freedom of association, legal impediments to the right of agricultural workers to organize remain in a significant number of countries, where national legislation either denies the right to organize in agriculture, or excludes the sector from the relevant legal protections. Some national governments may consider it impossible for their labour administrations to enforce this right in practice in rural areas. Others may consider that the nature of work in agriculture, with its atypical, seasonal or casual employment relationships, makes the sector less accessible to the right to organize.

International labour standards, because they are formulated and adopted by representatives of governments as well as employers' and workers' organizations, offer member States valuable guidance on what minimum standards should apply in the world of work. Not only do ILO Conventions and Recommendations stimulate improvements in national social and labour legislation, they also inspire good practices at the sectoral and enterprise level, as employers and workers integrate their principles into collective agreements.

MAJOR FACTOR INFLUENCING WORKERS' RIGHTS AND THE TERMS AND CONDITIONS OF EMPLOYMENT

A major factor influencing workers' rights and the terms and conditions of employment is their ability to bargain collectively with their employers rather than on an individual basis. When workers band together in trade unions and bargain together, *i.e.* collectively, their power is increased and they can secure better terms and conditions. Collective bargaining is a process of negotiation whereby terms and conditions of work are determined and rules created to govern labour relations. The participants in the process are employers and their organizations on the one hand, and worker representatives, usually from trade unions, on the other. The government is sometimes involved as a third party. The process usually culminates in the conclusion of an agreement, known as a "collective bargaining (labour) agreement (CBA)", though it may also be given other names. Individual bargaining between an employer and her or his employees is *not* regarded as collective bargaining.

Structurally, collective bargaining can take place at two levels: either at the enterprise (farm or plantation) level or the industrial or sectoral level. Issues which are collectively bargained can include wages, contracts of employment, labour contracting, maternity rights, health benefits, hours of work, leave, occupational health, safety and environment, housing conditions, grievance procedures, transport of workers, elimination of child labour, even measures to counter HIV/AIDS.

Given the predominance of women in the agricultural sector - notably in developing countries - and their vulnerable situation, it is particularly important that the gender implications of collective bargaining be understood and addressed. There are issues that in practice are of particular concern to women, which can be addressed through collective bargaining, such as equality of opportunity policies, equal pay for work of equal value, maternity leave and benefits, child care issues, reproductive health services. It is also important that the gender implications of apparently neutral issues for collective bargaining be assessed, including regarding wages, leave, overtime, bonus systems since these often in reality impact on women and men differently.

Collective Bargaining an Important Element in Improving Working

Collective bargaining is undoubtedly one of the principal instruments for regulating industrial relations. In many countries, however, collective bargaining does not appear to be a significant feature in the agricultural sector, largely because the respective institutions are lacking, governments do not encourage such negotiations, and trade unions and other forms of rural workers' organizations tend to be weak. It does not form part of the system of industrial relations. However, as more and more workers enter the rural labour market in search of wage employment, the importance of collective bargaining in the agricultural sector is increasing.

In general, CBAs are concluded in those sectors and enterprises where fulltime employment is significant and they tend to reflect the concerns of permanent workers. Other workers, whether daily, weekly, temporary or seasonal, may not be covered by such agreements, or may be only partially covered even if they form the majority of the total workforce.

Collective bargaining plays a more important role in plantation agriculture than in the sector as a whole. Because plantations are far better organized than the rest of the agricultural industry, the wages and terms of service set through collective bargaining have tended to be more favourable than those obtaining in the rest of the agricultural sector.

The right of workers to bargain collectively with their employers is supported by the ILO Right to Organize and Collective Bargaining Convention, 1949 (No.98), which is also linked to the right to freedom of association.

Contracts of Employment

Written contracts of employment tend to be rare in agriculture, where verbal agreements are commonly used. Such agreements tend to be rather general and align themselves to market or "common" conditions. Contracts of employment prevailing in agriculture display a great diversity, both within and between countries. Seasonal, daily and permanent contracts of employment can be found alongside task or piece rate employment. Permanent contracts are the least common form of contract and their share in total agricultural employment has been declining in most countries. The nature of contracts of employment can be linked to the skill requirements of the production process. Casual and seasonal migrant labour is often used in low-skill harvesting and processing operations in which labour costs relative to skills are a primary concern. The challenge is to find ways to balance the need for flexibility in hiring labour to meet production and marketing requirements with the basic standards and protection that should be inherent in any contract of employment. In the case of most casual and temporary forms of employment, this challenge has yet to be addressed.

EMPLOYMENT OF LABOUR THROUGH LABOUR CONTRACTORS

The employment of labour through labour contractors is a long-standing practice on plantations; it is also becoming increasingly common in other forms of commercial agriculture. Many agricultural undertakings retain a relatively small permanent core workforce whilst using labour contractors to source additional labour during peak periods such as harvest. Labour contracting occurs in many different scenarios within commercial agriculture. The nature of labour contractors themselves also varies; some are well known companies specializing in the recruitment, transport and management of waged agricultural workers whilst others are much less formal.

An essential characteristic of labour contracting arrangements is that the workers concerned do not have a direct employment relationship with the person or enterprise for which they perform work. In the majority of cases, the contracted workers remain the employees of the labour contractor. In some other cases, the workers may only be supplied to the workplace by the contractor and have no recognized employment relationship with either the user enterprise or the contractor. This gives rise to a number of practical and legal difficulties.

Where a worker is employed by a labour contractor and performs work for a user enterprise, the role and functions of an employer are actually shared between the contractor and the user. The contractor might pay the workers, provide transport to and from work and may in some cases provide workplace supervision. The user enterprise often determines the work to be done and the hours of work, and establishes the general terms and conditions of the worksite including in relation to occupational health and safety standards. This often means that the worker is uncertain about who her/his employer is, and thus about how to claim employment rights. These situations result in a lack of legal clarity around the employment relationship.

In some instances the system of labour contracting is abused. The provider of labour may not be a proper enterprise, but an intermediary of the supposed user enterprise, intended to conceal the user's identity as the real employer. This is a deliberate attempt to avoid the responsibilities of the employment relationship and as such is a fraudulent practice. Workers are particularly vulnerable in these situations; there are many well documented cases of the abuse of workers by so-called "labour contractors", such as asking for commissions, over-charging for transport, housing and food, holding back wages and imposing debt slavery.

Labour contracting is also occurring increasingly across borders in a variety of forms, some of which may be on the edge of legality or in the "grey zone" between the legal and illegal movement of persons.

A US study of agricultural injuries notes: "The increasing use of farm labour contractors represents an additional regulatory challenge for health and safety especially among migrant farm workers. The high degree of dependence of farm workers on the contractor for employment, transportation, housing and other amenities may hinder willingness to report work-related conditions." There are growing demands from trade unions for effective regulation of labour contractors and of the terms and conditions under which contract labour is hired and supplied.

An example is given in Part III, page 64. The ILO has been concerned with the vulnerability of workers in labour contracting arrangements for many years; this is reflected in the conclusions of successive ILO Tripartite Sectoral Meetings and an ongoing discussion at the International Labour Conference. A General Discussion on the Scope of the Employment Relationship was held in 2003 and a further discussion will be held in 2006 with a view to the preparation of an ILO Recommendation addressing issues relating to disguised employment.

Hours of work for waged agricultural workers tend to be long compared with other sectors, often over 45 hours per week, and remain largely unregulated. Hours tend to vary due to a variety of factors such as, for example, seasonal and climatic conditions, peak periods of sowing and harvesting. Adequate working time arrangements which can both limit overall daily, weekly and annual working time and meet the specific conditions of agricultural work cycles are, for the most part, still to be designed, negotiated and implemented.

An important breakthrough in this respect is Article 20 of the ILO Safety and Health in Agriculture Convention, 2001 (No.184) on hours of work which states: "Hours of work, night work and rest periods for workers in agriculture shall be in accordance with national laws and regulations or collective agreements".

This is the first time in an ILO Convention that the connection between hours of work, rest periods, night work, and health and safety on the job has been made, reflecting modern thinking and research on issues which were too often neglected in the past.

HOUSING AND LIVING CONDITIONS OF AGRICULTURAL WORKERS

Because many agricultural workers live where they work, their lives and occupations are inseparable. There is a close link between housing, worker well-being and productivity. Housing of agricultural workers is characterized by inadequate and overcrowded installations, no heating, poor ventilation, deficient sanitary facilities and non-potable drinking water, which enhance the spread of communicable diseases such as upper respiratory tract infections, influenza and tuberculosis. Living conditions on many farms and plantations remain inhuman, with workers living in tents, makeshift plastic huts/shacks or hostels for long periods. Poor housing also contributes to the spread of HIV/AIDS; such as male workers living in hostels away from their families, or families sharing overcrowded living quarters.

Examples of housing on plantations in Kenya illustrate some of the problems. Under Kenyan law, the companies have to provide housing for permanent staff. An ILO visit found that housing facilities in a number of plantations were in need of repair and upgrading so as to improve the living standards of workers, and some of the older houses required outright replacement. In some estates, workers were housed in rows of brick-built quarters with shared toilet facilities and water points, whereas in others, grass-thatched mud houses were still common. In some estates, workers complained of poor or non-existent sanitation, made worse by the congestion in the living quarters. In the lower-grade housing estates, one pit latrine was shared by several families, while drainage systems were non-existent in a majority of the cases. In some instances, employees had to draw water from nearby rivers and streams due to the absence of piped water. Cases of waterborne diseases were cited frequently among the workers living in such situations. Casual workers are rarely provided with plantation housing. Many are migrants who live in shanties near the estates without even the most basic sanitary facilities. The hygiene situation in these settlements can be deplorable, as witnessed by frequent outbreaks of waterborne diseases such as cholera, typhoid, malaria and dysentery.

Provision of decent housing on farms and plantations has been an issue, historically, on which trade unions have negotiated with farm and plantation employers. In some countries, employers are obliged by law to provide housing or a housing allowance to employees. However, when the accommodation is part of the remuneration package, this has often meant that workers are not free to change employers, as they would lose their housing. Also, they would lose their accommodation on retirement. Laws giving workers more secure housing rights on farm housing, including retirement provisions, have been introduced in a number of countries. Many unions, however, argue that wages should be sufficient to allow workers to buy on the housing market, like other citizens, and to live where they choose. Transport of workers In many countries, agricultural workers are transported over long distances on a daily or seasonal basis from their living quarters to places of work. All too often, large numbers of workers are packed in open trucks and vehicles never intended for the transport of human passengers. Weight limitations are disregarded and safety conditions ignored. There are many examples of serious road accidents involving such transport. It is often unclear whether such accidents are covered by employment injury insurance.

STANDARDS FOR AGRICULTURAL WORKERS AND SMALL FARMERS

Improving occupational health, safety and environmental (OHSE) standards for agricultural workers and small farmers, including regulation and enforcement, must be included as one of the key components of sustainable agriculture. Furthermore, there must be recognition that improving health, safety and environmental standards for workers can help protect and improve standards of public and environmental health, consumer food safety, and environmental protection, especially with regard to exposure to pesticides and other agrochemicals.

Agriculture is one of the three most dangerous occupations to work in, along with mining and construction. Waged agricultural workers face a wide range of hazards at work including, for example, dangerous machinery, unsafe electrical wiring and appliances, livestock-transmitted diseases, falls from heights, and exposure to toxic pesticides. The ILO estimates that there are some 355,000 on-the-job fatalities each year. It is estimated that half of them occur in agriculture, the sector with half of the world's workforce. Waged agricultural workers and farmers also suffer disproportionately among the estimated 270 million occupational accidents and 160 million occupational diseases each year. Each day, an average of 6,000 people die as a result of work-related accidents or diseases.

Agricultural work - and this is one of its most distinguishing characteristics - is carried out in a rural environment where there is no clear distinction between working and living conditions, unlike the case of the factory or office worker. As a result, agricultural workers and their families face extra dangers such as exposure to pesticides. Furthermore, the poverty experienced by waged agricultural workers and small farmers can contribute to increased risks of work-related ill-health due to bad diet and malnutrition.

Many of those killed, injured or made ill are women workers. They are especially at risk because they are often employed on a part-time or casual basis and receive less training and instruction, often do repetitive work which can result in musculoskeletal problems, and face reproductive hazards as a result of exposure to pesticides

Child labourers are also at risk – each year, 22,000 children are killed on the job, many of those in agriculture. An immediate priority is to eliminate all forms of hazardous work carried out by child labourers in agriculture. The type of employment relationship can also have a significant impact on health, safety and environmental practices. Kenyan trade unionists say subcontracting arrangements on plantations have made it difficult to tackle certain health and safety problems. For example, women weeders were regularly exposed to pesticides while at work, but there was little the union could do since the subcontractors who engaged the women were not bound by the collective agreement with the plantation company.

Human suffering cannot be assigned a monetary value, but the economic losses associated with poor occupational health, safety and environmental standards can.

High levels of deaths, accidents and ill-health resulting in lost workdays and increased medical expenses have a considerable negative impact on agricultural productivity and place enormous social and financial burdens on enterprises.

The ILO estimates that four per cent of Gross Domestic Product (GDP) is lost due to fatalities, accidents and work-related diseases. A safe, healthy and environmentally conscious workforce is an essential element of a profitable and sustainable agricultural industry. For trade unions, workplace organization and representation are the keys to maintaining and improving health, safety and environmental standards in the workplace. Legally appointed and empowered worker health and safety representatives (safety reps) are the backbone of trade union organization on health and safety at work. They are the eyes and the ears of trade unions on workplace occupational health and safety problems and play crucial roles in reducing fatalities, accidents and ill-health at work. Many deal with workplace environmental issues as well. These safety reps help protect workers, the public and the general environment.

However, on many farms there are no safety reps due to the small scale of these enterprises, their scattered nature, and a general lack of resources and technical support. Even on larger farms and plantations, where safety reps are selected or elected by the workers, or the legal right to do so clearly exists, safety reps may only have limited scope to improve occupational health, safety and environmental standards, due to a lack of training and technical support.

Article 8.1(b) of the ILO Safety and Health in Agriculture Convention, 2001 (No.184) gives workers the right to select their own workplace safety representatives. However, to turn this right into reality, what are termed "roving safety representative schemes" need to be put in place, based on external worker representatives who visit agricultural undertakings in a given geographical area.

Worker representatives on legally-constituted, joint worker-management health and safety workplace committees (safety committees) also play a vital role in maintaining OHS and even environmental standards. Such committees are only found in larger workplaces (generally, those with 50 employees or over). Safety committees deal with ongoing problems, and look at longer-term solutions on a joint problem-solving basis.

ADMINISTER SOCIAL AND LABOUR POLICY

To provide advice to both employers and workers, to administer social and labour policy, and to supervise and enforce labour legislation and standards, effective national systems of labour inspection are required. Labour inspection is a public function, a responsibility of government, best organized as a system, within the context of a larger state system. The ILO Labour Inspection Convention, 1947 (No.81) sets out the basic international standards, supplemented by the ILO Labour Inspection (Agriculture) Convention, 1969 (No.129) which take into account the special characteristics of the agricultural sector. Labour inspectorates work to secure the enforcement of the legal provisions relating to conditions of work and the protection of workers while engaged in their work, such as provisions relating to hours, wages, weekly rest and holidays, safety, health and welfare, the employment of women, children and young persons and other connected matters. Inspectors also supply technical advice and information to employers and workers concerning the most effective means of complying with the legal provisions. Labour inspection systems vary, with some countries having, for example, specialized health and safety inspectorates.

Labour inspectors in many countries also have an important role in overseeing the observance of trade union rights, the protection of legitimate workers' representatives (*e.g.* safety reps), and the effective operation of statutory bodies designed to engage in social dialogue. Labour inspectors in many countries are also charged with the task of providing training for worker representatives on all issues related to labour protection. If the inspectorate does not organize this training itself, inspectors will regularly participate in training activities organized by other parties. In many countries labour inspection, however, appears to be a low priority. The difficulty of effectively organizing a minimum presence of labour inspectors in large territories which lack resources cannot be underestimated. Many countries have too few labour inspectors for the adequate discharge of their duties in all economic sectors.

Labour inspectorates in agriculture in developing countries face special difficulties. First, even where legislation applies to agriculture, there are too few inspectors in these countries to ensure even a token appearance at more than a tiny proportion of workplaces. The larger the plantation or the more remote the agricultural activity, the greater the problem. Second, agricultural inspectors in these countries often receive inadequate training. Third, many inspectors complain of inadequate resources, particularly for transport, so that they are unable to travel to more remote farms and plantations.

THE HIV AND AIDS EPIDEMIC

AIDS (acquired immunodeficiency syndrome) is a disease caused by destruction of the immune system by a virus called HIV (human immunodeficiency virus). HIV and AIDS is a workplace issue, not only because it affects the workforce, but also because workplace health, safety and environmental training on this condition can play a vital role in limiting the spread and effects of the epidemic.

In 2004, some 37.8 million people are living with HIV; AIDS killed 2.9 million in 2003 (UNAIDS 2004). According to the FAO, 95 per cent of people living with HIV - and dying of AIDS - are in developing countries.

The overwhelming majority are the rural poor, and among them women figure disproportionately. The 2004 UNAIDS Reportxx estimates that the proportion of HIV-affected women has steadily increased. Latest figures show that in Africa there are 13 infected women for 10 infected men.

The epidemic is undoing decades of economic and social development and causing rural disintegration. For example, in sub- Saharan Africa, HIV and AIDS is depleting the region of its food producers and farmers, decimating the agricultural labour force for generations to come. AIDS has killed around 7 million agricultural workers since 1985 in the 25 hardest-hit countries in Africa, while the most affected African countries could lose up to 26 per cent of their agricultural labour force within two decades.

As the disease hits at the most productive age groups generally, most of the 40 million people infected with HIV are in the prime of their working lives. At least 25 million of those infected with HIV are workers aged 15 to 49. The effects are momentous - on workers and their families, enterprises, and national and regional economies.

The impacts on rural populations, their livelihoods, their farming systems, and on food security have been especially severe. The disease affects the world of work in many ways:

Households

- growing burden on women workers as they have to earn a livelihood while at the same time caring for sick family members and even neighbours;
- in agricultural communities, increasing number of households headed by children, as the parents have died from the virus and extended family networks cannot always cope with the sheer numbers of orphans;
- increasing pressure for AIDS orphans to be allowed to work in agriculture to cover the costs of their remaining on the farm or plantation and to pay school fees. However, there is a very real danger that these children will be exploited and their health put further at risk by exposure to occupational health and safety hazards;
- rural communities bearing a higher burden of the cost of HIV and AIDS as many urban dwellers and migrant labourers return to their villages when they become sick;

Workforce

- cutting the supply of labour and reducing income for many workers;
- discrimination against people with HIV, often resulting in dismissal;
- other ways in which HIV and AIDS affects rural workers include through changes in their ability to access and control productive assets and resources.

Farms/Production Units

- loss of valuable work skills and experience;
- falling productivity in industry and agriculture, and rising labour costs due to labour shortages.

Evidence indicates that vulnerability to HIV and AIDS is rooted in the social, economic, political and cultural landscapes. One of the most important factors of vulnerability is gender. HIV and AIDS is also a workplace issue in agricultural estates: research has shown that poor housing and living conditions play an important part in the transmission of the disease. A family or group of families may share one room, and in certain instances spouses are not allowed to live together on the premises. Seasonal labour migration results in a high concentration of men living in labour camps without their families. This was said to contribute to the habit of extra-marital sex (mostly by male spouses). All these practices coupled with poor working conditions and low wages have accelerated the spread of the disease among the workers. Access to information and health services is poor and people are unlikely to know how to protect themselves from infection. Poor conditions of work and low wages have driven many workers into behavioural patterns that increase the risk of infection and transmission of the disease such as involvement in commercial sex, sexual favours etc. The ILO has issued a Code of Practice on HIV/AIDS and the world of work. The Code states that governments "should ensure coherence in national HIV/AIDS strategy and programmes, recognizing the importance of including the world of work in national plans..." Employers, meanwhile, "should consult with workers and their representatives to develop and implement an appropriate policy for their workplace, designed to prevent the spread of the infection and protect workers from discrimination related to HIV/AIDS". In addition, an education and training manual on the Code has also been published.

HAZARDOUS NATURE OF AGRICULTURAL OCCUPATIONS

Due to the hazardous nature of agricultural occupations and the often remote location of farms and plantations, medical care is a critical employment benefit for agricultural workers. Through collective bargaining, trade unions, especially in plantations, have been able to negotiate with employers for better health care for their members than would otherwise be available. However, lack of public health facilities in rural areas is an aspect of sustainable agriculture which needs to be addressed. At the global level, health conditions in developing countries have improved spectacularly over the last 40 years.

However, the WHO estimates that around one billion poor people have been excluded from the global health revolution and would benefit from development programmes. Agricultural waged workers make up a significant percentage of that billion. The children of agricultural waged workers will also often be especially vulnerable to the childhood diseases of poverty, such as diarrhoea, acute respiratory infections, malaria, measles and perinatal conditions. Most if not all of these diseases are closely linked not only to poverty but also to lack of sustainable development in rural areas.

Furthermore, three billion people, over half the world's population, lack access to proper sanitation, one billion lack access to safe drinking water and 40 per cent of the world's population face water shortages. Over 800 million people who live in rural areas lack access to adequate and safe water.

GENDER AND MEDICAL OR HEALTH CARE

The link between gender, poverty and health is equally strong. Seventy per cent of the 1.2 billion people living in poverty in 2000 were female. Estimates over a 20-year period indicate that the increase in the numbers of poor rural women in 41 developing countries was 17 per cent higher than the increase in poor men. Protein energy malnutrition is significantly higher in women in South Asia, where almost 50 per cent of the world's undernourished reside. Half a million women die unnecessarily from pregnancy-related complications which are exacerbated by issues of poverty and remoteness. Almost all of these problems link directly or indirectly to the lack of sustainable agricultural development and the marginalization and vulnerability of women working in agriculture.

On plantations, employers have traditionally provided basic medical care in the form of dispensaries or clinics which are able to treat minor illness or injury. However, very little health and education data specific to plantations is available to establish any measure of progress.

Furthermore, lack of essential medicines is also a common problem in many countries. The WHO estimates that over a third of the world's population lacks access to essential medicines. Many of those concerned live in rural areas.

Lack of hygiene facilities, especially when working in the fields, is another problem. Workers and often their children are at risk. Parents often bring their children to the fields because there is a lack of day-care services for children in rural areas.

Therefore infants and young children are exposed to some of the same occupational hazards as their parents, even if they are not working. Workers and children are also at high risk of contracting parasitic and other infectious diseases because of the poor sanitary conditions in the fields and in their housing facilities.

CONTRIBUTION-BASED SOCIAL INSURANCE SCHEMES AND TAX-FINANCED SOCIAL ASSISTANCE

Social security is defined as being composed of contribution-based social insurance schemes and tax-financed social assistance. In their most comprehensive form, social security systems aim at providing nine types of benefit, namely, medical care, sickness and maternity benefits, family benefits, unemployment benefits, employment injury, invalidity and survivors' benefits, and old age benefits. One of the key global problems facing social security now is the fact that more than half of the world's population, workers and their dependents, are excluded from any type of social security protection. The problem is particularly acute in agriculture. ILO data show that fewer than 20 per cent of the world's agricultural workers are covered by one or more of the nine standard contingencies. As high levels of poverty and income fluctuations are characteristics of waged workers in agriculture, they are especially vulnerable economically when loss of wage-earning power occurs in event of death, injury, ill health, invalidity or natural disasters.

Even where there is assured legal coverage for agricultural workers, many social security programmes fail to reach all or some of the target group, and some potential beneficiaries fail to make good use of the benefits available. Few developing countries are able to provide comprehensive social security coverage with regard either to the contingencies or to the population covered. For these countries, improving social security protection is a key development objective.

For rural populations, the standards contained in the ILO Safety and Health in Agriculture Convention, 2001 (No.184) represent an important step forward. Article 21 states:

- In accordance with national law and practice, workers in agriculture shall be covered by an insurance or social security scheme against fatal and non-fatal occupational injuries and diseases, as well as against invalidity and other work-related health risks, providing coverage at least equivalent to that enjoyed by workers in other sectors.
- Such schemes may either be part of a national scheme or take any other appropriate form consistent with national law and practice.

CONTRIBUTIONS OF AGRICULTURAL WORKERS

The future contributions of waged agricultural workers and their trade unions to sustainable agricultural production, food security and sustainable development are potentially very large and virtually untapped. Part III provides examples of the contributions that waged agricultural workers are already making – both as workers and as citizens – to sustainable agricultural production, world food security and sustainable development.

It describes how their trade unions are supporting their efforts and seeking to increase their contributions to civil society in this area.

The examples demonstrate how workers and trade unions are meeting the new challenges by interacting, cooperating, and sometimes creating partnerships, with employers, governments, agricultural and rural development agencies, intergovernmental organizations, science and research institutions, agricultural banks and credit institutions and civil society organizations and groups.

Contribute to Sustainable Agricultural Production

Trade unions can use their workplace representation and structures, organizational abilities, experience of collective bargaining, technical know-how and education and training programmes, as well as their political commitment, to enhance their contributions to the achievement of sustainable agricultural production, food security and sustainable development. Ways in which they can do this include:

- launching union campaigns at workplace, community, local, national, regional and international levels;
- using union structures, institutional capacities and networks to enhance their contributions to sustainable agricultural production, food security and sustainable development issues;
- using union branch organizations and committees to raise worker awareness of, and commitment to, sustainable agricultural production, food security and sustainable development;
- negotiating "sustainability clauses" as part of collective bargaining agreements with farm and plantation employers. Workers have a vested interest in the long-term profitability and sustainability of the enterprises in which they work;
- ensuring that worker safety representatives also deal with environmental issues;
- encouraging joint worker-management health and safety workplace committees to also address environmental, sustainable agricultural production, food security and sustainable development issues;
- adapting union education and training programmes for grassroots members, shop stewards, safety representatives, safety committee members, and union officials to incorporate training on sustainable agricultural production, food security and sustainable development. At the moment, these are "buzzwords" which do not mean anything

to the vast majority of grassroots agricultural workers and even leaders;

- using national tripartite structures and procedures - involving government, employers and trade unions - to promote these issues. Also, making use of parliamentary and legislative processes to ensure, for example, policies and standards on these issues;
- encouraging inter-regional cooperation between unions, and promoting these issues in appropriate regional fora and processes;
- encouraging international cooperation between unions, and promoting these issues in appropriate international fora and processes;
- strengthening links between agricultural workers and their families and local communities on these issues;
- networking and alliance building with non-governmental organizations and civil society organizations;
- building links and alliances with agricultural producer organizations.

Trade Unions and Multinational Enterprises

A framework agreement is an overarching agreement directly negotiated between a company and a trade union on labour rights, standards and employment throughout all parts of the company's operations. In 2001, the IUF and the Coordination of Latin America Banana Workers' Unions (COLSIBA), for example, signed a framework agreement with Chiquita, a multinational banana company. According to the agreement, Chiquita will respect core labour standards, and will require its suppliers, contract growers and joint venture partners to respect these same standards. There is also provision for joint evaluation and monitoring to ensure that labour standards are being met.

Supply Chains in Agriculture, Codes

Trade union work on supply chains in agriculture, Codes of Conduct, fair trade labelling and ethical trading initiatives includes:

Over the past decade or so there has been a substantial growth in work on corporate social responsibility. Initially, this interest was driven by civil society organizations (CSOs) concerned about environmental and child labour issues. More recently, this has expanded to a wider interest in working and social conditions in producer countries. Many companies responded to these concerns by issuing unilateral codes of conduct which often had little basis in accepted international norms, *e.g.* International Labour Standards, and which were little more than publicity stunts.

The international trade union movement responded by issuing a model code of conduct based firmly on international labour standards, which trade unions could use to challenge companies and to negotiate more genuine codes. However, many international trade unions, including the IUF, prefer to enter into direct negotiations with the companies through framework agreements. The cut flower sector is characterized by extensive use of (young) women workers on seasonal contracts, heavy use of pesticides, problems of waste disposal and high water usage. Non-governmental organizations (NGOs) in the North were quick to raise concerns about flower production - especially but not exclusively the environmental costs. The flower industry responded by issuing a plethora of unilateral codes of conduct.

To address the problem of unilateral codes, the IUF, working with affiliates and several NGOs in a body called the International Flower Co-ordination drew up a model International Code of Conduct for the Production of Cut Flowers. This Code is based firmly on International Labour Organization standards. Importers, especially in Germany and the foundation that sets the environmental standards for the Netherlands-based flower auction, were targeted to convince them to accept the International Code of Conduct. Workshops were held for East African trade unions on the International Code of Conduct and how to use it to organize workers and to improve their working conditions. A Training Manual for shop stewards on how to use the Code has been developed. Further negotiations with the flower producers have led to the introduction of a Fair Trade in Flowers and Plants scheme coordinated by an industry body, Union Fleurs.

The work around promotion of the International Code of Conduct for the Production of Cut Flowers has highlighted the many problems that IUF affiliates have with codes of conduct, even those drawn up multilaterally and based on ILO standards. The evidence so far is that it is very hard for trade unions in producer countries to use codes of conduct to improve working conditions. To date, there are no examples of a code, even with freedom of association as its cornerstone, leading to the formation of a new union. Some unions have been able to use the code to establish new branches but this is still fairly exceptional and there are some examples of improvements in living and working conditions, especially when a union has been able to append the code to its collective bargaining agreement.

Ethical Trading Initiative

The Ethical Trading Initiative (ETI) is a UK-based membership organization of companies, trade unions and NGOs which was established in 1998 to identify and promote good practices in the implementation of codes of conduct. The ETI has its own base code negotiated by all members and based on International Labour Standards. All stakeholders agree that the ETI base code was developed as a tool to achieve positive change on the ground.

The ETI aims to enhance the private sector's contribution to sustainable development by encouraging business practices that embrace social, environmental and financial responsibility. Ethical supply chain management is a critical aspect of responsible business in developing countries. The IUF is participating in the ETI at board level and has also been involved in pilots in the agricultural sector, *e.g.* horticultural products and bananas.

"Fair-trade is a trading partnership, based on dialogue, transparency and respect, which seeks greater equity in international trade. It contributes to sustainable development by offering better trading conditions to, as securing the rights of, marginalised producers and workers – especially in the South".. "Fair-trade" is therefore a recognized term for agreements between producers in developing countries and commercial buyers who wish to purchase and market products based on stable and "just" or "fair" prices and production criteria which respect labour and environmental standards. Fair trade aims to increase producers' access to markets, improve their incomes, and ensure that their production is based on sustainable development principles.

The Fair-trade Labelling Organizations International (FLO), for example, sets common criteria for fair trade tea, coffee, cocoa, honey, orange juice and bananas. FLO works mainly to label goods from small farmers, but in the tea and banana sectors there are also plantations, and the IUF's concern has been to understand fair trade's impact on employed workers and how fair trade can help them both in organizing trade unions and in improving living conditions, without undermining collective bargaining.

INTEGRATED PRODUCTION AND PEST MANAGEMENT IN AGRICULTURAL WORKERS

Integrated Production and Pest Management (IPPM) is a way of growing crops that maximizes control of pests by their natural enemies - pests, parasites and pathogens (diseases), integrated with other crop husbandry measures. This management technique aims to keep pest populations below economically damaging levels and to restrict pesticide use to amounts that are economically justified and reduce risks to human health and the environment.

Agricultural workers often say, "We know that chemical pesticides are bad for our health and that of our families and communities. So what are the alternatives? How do we stop using these poisons?"

One answer is to ensure workers are trained to understand and use IPPM techniques. Normally, it is only farmers who receive IPPM training, especially through an educational method called "Farmer Field Schools" (FFS). The FAO has been promoting the use of such techniques through farmer field schools in its country programmes throughout the world.

The FAO's integrated pest management programme (IPM) has been particularly successful in Asia and in 1993, the FAO inter-country programme on IPM rice in Asia organized a global IPM meeting to introduce its successful IPM approach to interested policy makers from other regions. Consequently, the FAO, World Bank, United Nations Development Programme and United

Nations Environmental Programme established the Global Integrated Pest Management Facility in 1995. This joint programme is housed in the FAO and is the main international agency promoting IPPM worldwide.

The IUF is now working with the Global IPM Facility to train agricultural workers in IPPM techniques, using the FFS method. Field Schools mean that workers, like farmers, go into a field to study how a crop grows, to learn to identify harmful insects, diseases and weeds, and to learn to identify how to protect and encourage beneficial insects. The workers then draw up their own agro-ecology plan for that particular crop and field, setting out how to grow a healthy crop and how to protect it from pest and disease attack and weed competition by non-chemical means.

Equipped with this new knowledge, workers can then negotiate clauses requiring use of IPPM programmes in collective bargaining agreements with employers. The aim is to give agricultural workers knowledge and skills on IPPM so that when instructed by an employer or manager to use a toxic pesticide, they can point out that IPPM techniques provide a safer way of controlling the weed, insects or diseases. Safer for themselves and the supervisors, the managers, the community and the environment and for the crop (which may then be sold at a premium price).

Pilot IPPM courses - the first ever of their kind for waged agricultural workers - were held in 2001 for agricultural trade unions in Tanzania (TPAWU) and Uganda (NUPAW and NUCMAW). The unions concerned also invited some NGOs and organic farmers' organizations to join the courses. Training was given by professional IPPM trainers provided by the Facility. The pilot training is ongoing, with a view to expanding it to other unions and countries.

Pesticide Management Practices

To improve workplace occupational health, safety and environmental standards, especially targeting fatalities, poisoning, ill-health and pollution resulting from intensive pesticide use, the IUF started an Occupational Health, Safety and Environment Project in 1998.

The Project aims to build the capacities of affiliated national unions and the IUF's regional and international networks to tackle occupational hazards within the context of promoting integrated production and pest management and sustainable agriculture. On Health, Safety and Environment, A Series of Trade Union Education Manuals for Agricultural Workers have been developed by the IUF and ILO, which are also designed for use by small farmers and non-governmental organizations.

Training targets grassroots members, worker health and safety representatives, worker representatives on joint worker-management health and safety workplace committees, and branch and national union officials and committee members.

To date, seven participating African agricultural unions in Ghana, Malawi, Tanzania and Uganda have each trained a core of trainers from among their members. These trainers run regular training courses for the target groups working in cocoa, coffee, flower, horticultural, sugar, tea, and tobacco production. Results of the training include improvements in workplace OHSE standards, recruitment of new members, and the unions having new influence at national level on OHSE laws and policies.

The Project also supports the IUF's campaign with its national affiliates to ensure ratification and implementation by governments of ILO Convention No 184 on safety and health in agriculture.

There is also close cooperation at local, regional and international levels with the Pesticides Action Network (PAN). In Latin America, IUF and PAN have signed a cooperation agreement and are working together, for example, to ban toxic pesticides. In East Africa, the IUF's agricultural affiliate unions in Tanzania and Uganda are working to help develop the PAN network in other anglophone African countries.

At the international level, the FAO International Code of Conduct on the Distribution and Use of Pesticides, adopted by the FAO Conference in 1985, identified the shared responsibilities of governments, industry, users and international and non-governmental organizations in all aspects of pesticide distribution and use, and formulated voluntary standards of conduct for each party. The Code is highly relevant to farmers and agricultural workers in developing countries since they use pesticides in their farming activities.

In 1989, the principle of Prior Informed Consent (PIC) procedure was included in the Code, whereby international shipment of a chemical that is banned or severely restricted (in order to protect human health or the environment) should not proceed *i.* without the agreement, where such an agreement exists, or *ii.* contrary to the decision of the relevant authority in the importing country. The IUF was active in helping in the 2000 revision of the Code.

The principle of PIC is now incorporated in the UN's Rotterdam Convention on the Prior Informed Consent (PIC) Procedure for Certain Hazardous Chemicals and Pesticides in International Trade which is jointly administered by the FAO for pesticides, and the United Nations Programme for the Environment (UNEP) for industrial chemicals.

The IUF represented the trade unions internationally in the negotiations to develop the Rotterdam Convention and continues to participate in the meetings of the governing body for the Convention, as well as promoting its ratification and implementation at national level.

The IUF has also been active in helping set up and participating in the work of the Intergovernmental Forum on Chemical Safety (IFCS), in which both the FAO and the ILO actively participate. The Forum is a body where governments, intergovernmental organizations, industry, scientific bodies, trade unions and public-interest NGOs meet to:

- Develop and coordinate policies on the environmentally sound management of toxic chemicals
- Provide advice and make recommendations to governments, intergovernmental organizations, industry, trade unions, scientific bodies, and public-interest NGOs.

The IUF is also currently representing the international trade unions on the IFCS Steering Committee.

The IUF has also been active in another international initiative related to Agenda 21 to develop National Profiles to Assess the National Infrastructure for Management of Chemicals.

The IUF participated in the working group developing the format for the Profiles. IUF agricultural affiliate unions in Ghana, Tanzania and Uganda have participated in the development of their countries' National Profiles.

They are now helping to implement the resulting action plans, in order to improve national pesticide health and standards for the benefit of agricultural workers and farmers, local communities, the environment and consumers.

Industry Joint Training

The IUF's OHSE Project is working with Crop Life International (CLI), the international trade association of the transnational pesticide manufacturing companies, to jointly train agricultural workers - and even management - on pesticide hazards and risk management prevention and control measures. The IUF and CLI each provide one to two trainers per course, who then jointly train workers and management.

Pilot courses began in 2000 for two trade unions in Uganda, NUPAW and NUCMAW, and have continued on a regular basis since then. This cooperation arose in large part from an ILO ACTRAV-sponsored study in 2000 of the pesticide industry's voluntary "Safe Use" training project in Guatemala.

The IUF was especially critical that the industry project did not provide training for waged agricultural workers or their trade unions, and that the training needed to be based on solid education and training methods and less presentational in style.

Develop Roving Safety Representatives in Agriculture Schemes

Considering the close relationship between occupational health, safety and environmental standards and productivity, ensuring healthy and safe work practices through joint efforts is an area of potential cooperation between employers and trade unions. Much is to be gained in terms of workers' wellbeing and enterprise productivity.

UK

The United Kingdom's government agency, the Health and Safety Executive, stated publicly in 2001. "There is strong evidence that workplace-based safety representatives make a positive contribution to improving health and safety standards. We need to know whether similar expertise can help to improve employee consultation and health and safety at workplaces that currently do not enjoy the benefit of safety reps." To examine the effectiveness of trade union appointed roving safety representatives in agriculture, trained mainly by the trade union, compared with professional health and safety advisers, the Health and Safety Executive funded an 18-month research project. The project was initiated through the tripartite (Health and Safety Executive) Agricultural Industry Advisory Committee, and managed by the Agricultural Development Advisory Service (ADAS).

The agricultural workers' section of the Transport and General Workers Union (TGWU) fully participated in the project. They selected the eight volunteer worker roving safety reps and helped train them, in cooperation with ADAS. The employers' organization, the National Farmers Union, also cooperated in the project.

Around 100 volunteer farms were identified and an initial health and safety status review was conducted to establish a baseline. The farms were then divided into three groups.

One group of some 40 farms was visited by the eight worker roving safety reps, each of whom made some four visits to a given number of farms in their respective areas. The second group of 40 farms was visited by the professional advisers (ADAS staff). The remaining 20 farms were a control group and received no visits. At the end of the process, there was a final status review of the farms in all the groups.

The project found that roving safety representatives had proved to be effective. However, securing access to employees, an important criterion for encouraging greater worker participation in improving health and safety standards, proved problematic in some instances.

South Africa

An example of tripartite cooperation - government, employers and trade unions – is a project to develop a roving safety representatives scheme in South African agriculture, based on the system legally operating in Sweden. The project aims to develop safety rep schemes to cover small to mediumsized agricultural undertakings, and to boost the provision on larger farms and plantations.

Roving safety representatives have the same legal powers as ordinary safety reps except that they can visit farms where they themselves are not employed. Based on common goals shared by workers, trade unions and employers, roving safety representatives help raise awareness, provide information, and advise on problems at the workplace in order to prevent and reduce fatalities, accidents and ill health.

In June 2002, the Swedish trade union Kommunal, in cooperation with the IUF, funded a six-person South African tripartite delegation visit to study the Swedish roving safety representatives scheme. The visit by the government, agricultural trade union and agricultural employers' representatives aimed to (a) see the roving safety representatives scheme in action on Swedish farms; (b) promote understanding about the way the union-nominated roving safety reps works, and their legal rights and duties - in particular, the contribution of each of the stakeholders to the system; (c) learn about the training programmes for roving safety reps, what back-up support is needed and from whom, and how the system is financed; and (d) develop ideas about how the scheme - or parts of it - could be applied in South Africa. Tripartite discussions are now underway to set up pilot RSR schemes in three regions in South Africa, and Kommunal and the IUF are continuing to facilitate the process.

ENVIRONMENTALLY SUSTAINABLE DEVELOPMENT

Especially since the UN Conference on Environment and Development in 1992, trade unions around the world have become more active on environmental issues, and on networking and building alliances with other groups.

For example, the Labour Environmental Alliance Society (LEAS) in Vancouver, Canada, involves workers directly in environmental issues by linking workplace issues and the environment and making the connection between health and the environment. It also works to foster understanding of the labour movement among environmentalists. LEAS has an extensive network of over 50,000 workers, environmentalists and social justice activists who have found common ground.

The Society's main focus is on eliminating toxins from the workplace and community. It is especially concerned about the impact of pesticides in food and the effects on workers and consumers, notably children. It runs educational workshops with the British Columbia Federation of Labour and the Canadian Labour Congress as well as working with citizen, environmental and consumer groups. One of its member organizations, Farmfolk/Cityfolk, is a national leader on food issues. The Society is also involved with local and coastal fishers on issues of sustainable fisheries, food distribution and fish habitat, and water quality protection.

An international labour foundation promoting trade union interests in sustainable development, SUSTAINLABOUR, was established in October 2004 in cooperation with the Spanish government, with the participation of UNEP, ILO and major national and international trade union organizations working actively on sustainable development issues. Based in Spain, the objectives of the foundation are to examine major challenges and opportunities and problems facing trade unions in this field, and to develop networking, capacity-building and training programmes at the regional level. It will develop country profiles from a labour perspective through research and measurement, designing of national campaigns and corporate assessments. At the workplace level, it will develop methodologies for workplace assessments and training methodologies on clean development mechanisms. In particular, it will continue and reinforce work started by the International Trade Union Network on Sustainable Development (ICFTU/TUAC/ETUC) which culminated in the strategic participation of over 400 trade union representatives in the 2002 World Summit on Sustainable Development, on issues such as SARD good practices, clean water, sanitation and human settlements and other topics affecting workers throughout the world. In particular, SUSTAINLABOUR will follow up on sustainable development and the social dimensions of globalization. One of the major aims of the trade union foundation is to make effective input and to follow the current discussions and preparations for the Commission on Sustainable Development, which meets annually at the United Nations in New York.

Food Safety and Food Security

The IUF aims to promote food safety and food security, based on its historical and statutory commitment to "actively promote the organization of the world's food resources for the common good of the population as a whole." An example of trade union action on food safety, which aimed to protect both workers and consumers, was the Transport and General Workers Union (TGWU) involvement in the Bovine Spongiform Encephalopathy (BSE) crisis - the "mad cow" disease crisis - in the livestock industry in the UK in the 1990s.

In response to the crisis, the TGWU advised its members on the health and safety implications, particularly in relation to Creutzfeldt-Jakob disease and made considerable efforts to avoid job losses. The union also called for a public enquiry into the causes of BSE and the link with Creutzfeldt-Jakob disease, for the monitoring of food safety and hygiene to be moved from the Ministry of Agriculture, Fisheries and Food to the Department of Health and the Health and Safety Commission, and for independent research work to examine the risks of contracting Creutzfeldt-Jakob disease for workers in the farming and food industries. The union also made representations to the UK government and the European Commission on these issues.

The FAO clearly plays a key role in promoting food safety and quality internationally, but to date, there has been little international trade union involvement in this work. Improving FAO-trade union cooperation on food safety issues and their impact on food trade could be a potential area of work to be developed in the future.

THE SUSTAINABLE AGRICULTURE AND RURAL DEVELOPMENT INITIATIVE

In 2002, FAO's Director-General, with Governments and civil society Major Groups launched the SARD Initiative, a multi-stakeholder framework designed to support the transition to people-centered sustainable agriculture and rural development and to strengthen participation in programme and policy development. The ILO's Director-General, also expressed his organization's support for the Initiative.

The SARD initiative is led by civil society, supported by governments and inter-governmental agencies, and facilitated by FAO, in recognition of its role as Task Manager for SARD. At the WSSD, over 65 governments and civil society organizations from 90 countries expressed support for this initiative, with the expectation that FAO could be an appropriate and neutral convener.

In 2003, the SARD Initiative was selected as one of the Director General's twelve high visibility programmes and in 2005, in its 19th Session (April 2005), FAO's Committee on Agriculture *"agreed that the SARD Initiative is an important instrument for fulfilling FAO's responsibility as UN System Task* Manager for WSSD follow-up on implementation of Agenda 21" *and "welcomed FAO's continued support [...]"*, demonstrating that the Initiative responds to issues identified in FAO governing bodies as requiring support through FAO.

The Major Groups are those established by the Agenda 21 programme which was agreed by governments and stakeholders at the UN Conference on Environment and Development, 1992.

They are: agricultural workers and their trade unions, farmers, business and industry, non-governmental organizations, indigenous peoples, women, children and youth, scientific and technological community, non-governmental organizations, and local authorities.

The SARD Initiative was developed to move beyond simply cataloguing successful case studies and it is now intent on multiplying the number of successes by fostering interactions between major groups at all levels and the relevant institutions of governments and international non-governmental organisations, and international governmental organisations.

In order to help eradicate poverty, to reduce the vulnerability of rural people and to increase their self-reliance, the objectives of the SARD Initiative are to:

- Build the capacities of rural communities, organizations and networks to better access resources, promote fairer conditions of employment in agriculture, and adopt innovative practices for SARD
- Strengthen the cooperative relationships and work among Major Groups
- Contribute to the implementation of Agenda 21, as well as the SARD-related elements of the Millennium Development Goals and the World Summit on Sustainable Development Plan of Implementation.

The IUF is acting as the Major Group focal point for agricultural workers and their trade unions in the SARD Initiative. In particular, it is taking the lead on developing ideas and activities to promote fairer conditions of employment in agriculture, which is one of the three major thematic areas of the Initiative. This report is in part a contribution to this programme. The two other areas are: improving access to resources and promoting good SARD practices. FAO organized an Electronic Forum on Sustainable Agriculture and Rural Development (the SARD E-FORUM), between 24 June 2002 and 17 August 2002, in preparation for the World Summit on Sustainable Development, with the aim of promoting SARD and developing SARD implementation initiatives related to the following areas: Access to Resources; SARD Good Practices, Principles and Case Studies; and Fair Conditions for Employment in Agriculture. Of these, the discussion on fair conditions for employment in agriculture was closely linked to the contribution of agricultural workers to SARD and it aimed to solicit a variety of perspectives and insights on fair employment in agriculture and the role of waged agricultural workers in sustainable agriculture; build an experience base of successful cases related to improving the conditions of waged agricultural workers and their contributions to sustainable agriculture; and identify consensus points on how to achieve fair employment conditions for agricultural workers. The outputs included: consensus on leading issues regarding fair employment for agricultural workers; documentation of successes related to agricultural workers and sustainable agriculture and rural development; and key action areas for the future related to fair employment in agriculture.

A variety of topics were identified as crucial to addressing the issue of fair conditions of employment, including the question of child labour and young workers in agriculture, international standards, especially freedom of association, the right to collective bargaining, non-discrimination in employment, the abolition of forced labour, and the elimination of child labour, and specific ILO standards on agriculture that cover safety and health, labour inspection, rural workers' organizations and plantations; health, safety and environmental standards in agriculture; community-oriented models for employment; fair employment in agriculture and the WSSD (especially through promotion of policies that recognize employment as central to poverty reduction, and initiatives to increase labour productivity, the level of decency of work and income-generating capacity).

The SARD campaign on fair employment focuses on Major Group efforts (in cooperation with other stakeholders) to strengthen economic and social standards/conditions of SARD with respect to promoting fair conditions of employment in agriculture for permanent workers and temporary and/or seasonal (migrant) workers, wage-dependent smallholders, self-employed farmers working as contract labourers, self-employed workers employed as contract labourers, and sharecropping, through the improvement of health, safety and environmental standards for farmers and agricultural workers, and fairer conditions of employment for women in agriculture.

TRADE UNIONS REPRESENTING WAGED AGRICULTURAL WORKERS

Trade unions representing waged agricultural workers have taken a keen interest in efforts to achieve food security. In the run-up to the World Food Summit (WFS) in 1996, the 13th FAO/International Trade Union Consultation focused on food security, and how trade unions could contribute to the Summit and in the implementation of the WFS Plan of Action. The consultation recommended that trade unions participate in the "dialogue at the national, regional and global levels that has been launched on food security in preparation for the World Food Summit" and "provide information to their constituencies... contribute feedback to the Plan of Action and technical papers which were prepared for the Summit, and.........identify areas for cooperation in the Plan of Action".

The FAO took steps to promote the participation of trade unions in the policy dialogue on food security in the preparatory processes of the WFS, notably by preparing, in cooperation with trade unions and other organizations, the document *Partners for Food Security*. Intended as a contribution to the WFS process, this document brings together the trade unions' mandates, strategies and experiences in promoting food security, and their positions and views on the key issues addressed by the WFS and the Plan of Action. It presents possible areas for cooperation in implementing the WFS Plan of Action.

The trade unions expressed their readiness to bring the views and positions of their membership on food security issues to the attention of policy makers at country, regional and international levels. As an example, they pointed to the participation of trade unions and farmers' organizations in the Latin American and Caribbean Movement for Food Security and in drawing up the Managua Declaration of 1995 in which governments, people's organizations and NGOs pledged to work together on strategies for achieving food security in the region. Some trade unions indicated their capacity to engage in research on food security issues and to act as channels for the dissemination of research and information.

The trade unions also indicated that cooperation at local and national levels between governments, cooperatives, trade unions and other sectors of civil society is key for meeting the needs of the food insecure. Areas for cooperation include: provision of social welfare and safety nets and integrated programmes for sustainable agriculture and rural development. The IUF participated in both the 1996 and the 2002 World Food Summits. It has also acted as the focal point for trade unions in the International Non-governmental Organizations (NGO)/

Civil Society Organizations (CSO) Planning Committee on Food Sovereignty and Food Security. This body is coordinating and promoting NGO/CSO follow-up to the World Food Summit: Five Years Later, 2002. The IUF has written a section on workers' rights for the proposed International Code of Conduct on the Right to Adequate Food which the NGOs/CSOs put forward as one of the main outcomes of the Summit, and on which the FAO has now set up a working group.

ELIMINATION OF CHILD LABOUR

Trade unions have conducted campaigns, programmes and activities to promote decent work in agriculture and to use their collective strength, through either collective bargaining or grassroots action, to eliminate the use of child labour. The aim is to ensure that children have a chance to live in a safe environment, particularly one free from hazardous work, and to have access to school, so that they can achieve their full potential. Trade union work to eliminate use of child labourers in agriculture includes:

Sectoral Level

The IUF has signed an agreement with the International Tobacco Growers Association to work together to eliminate child labour in the tobacco industry. The IUF's training and education programmes have enabled affiliates to secure collective bargaining agreements which commit employers to eliminate the use of child labour. A Foundation for the Elimination of Child Labour in Tobacco has now been set up, comprising tobacco manufacturers and trade unions.

In 2001, following media revelations about the use of child slave labour in cocoa production in some West African states, the IUF and several NGOs began talks with the main companies and organizations in the global cocoa and chocolate industry. The discussions resulted in the establishment in 2003 of the International Cocoa Initiative: Working towards responsible labour standards for cocoa growing, based on a joint Foundation like the one in the tobacco sector. The Foundation will focus initially on the elimination of child labour in West African cocoa production.

National Level

The Brazilian agricultural workers' union, CONTAG, has been at the forefront of the fight to stop child labour. In 1993, in partnership with the ILO's International Programme for the Elimination of Child Labour (IPEC), the union started a "Child Workers' Action Programme" in 88 municipalities where there were large numbers of rural workers.

The Action Programme produced 10,000 copies of a booklet on the rights of rural working children and provided five training courses for 150 union leaders and monitors on how to improve provisions in collective bargaining agreements on the prohibition of child labour. Seven highly successful radio programmes aimed at awareness-raising were produced and broadcast on CONTAG's network of 160 local radio programmes.

In rural areas in Kyrgyzstan, almost 100 per cent of children work before and after school in the fields. Low prices for agricultural products mean that small farmers cannot afford to hire adult labour. A joint project of the ILO Bureau for Workers' Activities (ACTRAV) and the IUF is training trade union representatives in rural areas on child labour issues with a view to helping eliminate the most hazardous forms of child labour in cotton, rice and tobacco production. The Kyrgyzstan Agricultural Workers Union, an IUF affiliate, has well-developed infrastructures in all regions of the country which helps with project implementation.

The project also aims to help small farmers to increase productivity and income so that they will be able to hire adult workers - a measure which will also help reduce rural unemployment and migration. Furthermore, the union owns small plots of land in all seven regions, which it uses to train small farmers. It also uses its land as a guarantee (collateral) to secure bank credits for farmers who wish to convert to more sustainable forms of agricultural production and for developing micro-credit cooperatives for union members. In Ghana, the General Agricultural Workers Union has been able to negotiate a collective bargaining agreement with the Ghana Oil Palm Development Company which contains the following clauses: "The management is committed to the eradication of child labour in and around the plantations, and within the country as a whole" and "The management shall, in conjunction with the Union take necessary action to ensure that child labour is absent from within and around the plantation".

TRADE UNIONS AND AGRICULTURAL PRODUCERS' ORGANIZATIONS

One of the key questions that agricultural trade unions have addressed from the 1990s onwards, especially in the light of the impacts of globalization and structural adjustment policies, is how to strengthen relationships between trade unions representing waged workers and small farmer organizations. Whilst there are many common issues that could unite the two groups, the relationships and interaction between waged agricultural workers and small farmers can pose dilemmas for trade unions. What happens, for example, when small farmers are employers? Equally, in the process of land reform, workers and farmers can end up fighting each other, especially in the absence of any arbitration system. Farmers take over the land and workers – often migrant workers, who have no right to stay in the country – lose their livelihoods. To build links and alliances between trade unions and small farmers' organizations, the IUF started a Land and Freedom Project in 1998.146 The Project aims, for example, to ensure the common defence of legal rights, improved access to land, and greater influence on local, national, regional and international policies, and to counter the concentration of power of multinational enterprises in the agricultural sector. It is developing models to help trade unions and small farmers to work more closely together, and to adapt trade union structures, rules and training programmes so unions can recruit small farmers as members.

In 1994, in the framework of the FAO's collaboration with international trade unions in support of the activities of rural workers' organizations in sustainable rural development, the FAO undertook a study on the General Agricultural Workers Union (GAWU) of Ghana. The report focused on the organization of the self-employed rural workers and the rural development role of agricultural unions. A special focus of the union's activity was building partnerships with other organizations working in the rural areas and promoting the participation of women in trade unions, within the more general objective of increasing workers' participation in rural development. GAWU was one of the first agricultural unions to address the problem of the decline in union membership in rural areas, and it proposed the strategy of establishing partnerships with other community-based organizations working locally. GAWU also aimed at opening its membership to all rural workers, including both wage-earners and the self-employed, so as to extend the role of trade unions beyond the worker-employer relationship and pursue broader goals such as agrarian reform and rural development. GAWU has played a rural development role through different interventions, such as support services targeted at the self-employed, group enterprise promotion, transfer of improved technology and techniques, economic support services and social and community development activities. However, the main challenges that GAWU encountered in its strategy were in the phases of the operationalization of its programmes.

Building networks and alliances between waged agricultural workers and small farmers, and the trade unions and producers' organizations that represent them, is key to ensuring the participation of these groups and organizations in national decision-making processes, and in achieving sustainable agriculture and rural development. Increasing participation of workers and small farmers in these processes, and strengthening the institutions that represent them, can facilitate the creation of employment opportunities and access to employment, and help improve working conditions.

Bibliography

A C Mittal : *Energy and Sustainable Development*, Vista International, Delhi, 2008.

A.K. Shrivastava : *Encyclopaedia of Environmental Pollution, Agriculture and Health Hazards (5 Vols-Set)*, APH Publication, Delhi, 2004.

Ajit K. Ghosh.: *A Modern Book on Plant Biotechnology*, Cyber Tech Publication, New Delhi, 2009.

Amitava Rakshit and Priyankar Raha.: *Terminologies in Agricultural Sciences*, IBDC Publication, Delhi, 2009.

Ashok Kumar Sharma : *A Textbook of Biochemistry*, Random Publication, Delhi, 2012.

B Balaswamy : *Communication for Sustainable Development*, Concept Publisher, Delhi, 2006.

Bidhan Roy and Asit Kumar Basu.: *Abiotic Stress Tolerance in Crop Plants : Breeding and Biotechnology*, New India Publication, Delhi, 2009.

Brajendra, Kailash Chand Verma, Gyan Prakash Mishra, P. Muthuraman.: *Objective Agricultural Science*, New India Publication, Delhi, 2009.

Bruno Dorin and Thomas Jullien : *Agricultural Incentives in India* : Challenges, Oxford University Press, Delhi, 2005.

C.P. Malik, Bhavneet Kaur and Chitra Wadhwani.: *Advanced Topics in Biotechnology and Plant Biology*, MD Publication, Delhi, 2008.

C.R. Sudharmai Devi : *Analytical Procedures in Soil Science and Agricultural Chemistry*, Agrotech Publisher, Delhi, 2004.

Dewasish Choudhary : *Basics of Agricultural Chemistry*, Anmol Publication, Delhi, 2009.

G Byju : *Soil Chemistry : Basic Concepts*, New India Publication, Delhi, 2008.

Harry H. Love : *Application of Statistical Methods to Agricultural Research*, Cosmo Publication, Delhi, 2006.

J.B. Ambekar : *Agricultural Communication and Sustainable Development*, Shruti Publication, Delhi, 2010.

Jennifer A. Elliott : *An Introduction to Sustainable Development*, Routledge Publsher, Delhi, 2010.

K.S. Dadhwal and Mohit Gahlot : *Advances in Agricultural Research in India*, International Book Distribut, Delhil, 2002.

Kazif Jalal and John A Boyd : *An Introduction to Sustainable Development*, Prentice-Hall of India, Delhi, 2007.

N.K. Prasad : *Soil Fertility and Plant Nutrition, IBDC Publishers*, Delhi, 2013.

P N Mishra : *Environment Pollution and Its Control*, Sumit Enterprises, Delhi, 2007.

R.A.S. Tomar : *Encyclopaedia of Agricultural Chemistry*, Anmol Publication, Delhi, 2010.

Shivanand Tolanur : *Practical Soil Science and Agricultural Chemistry*, International Book Distributing, Delhi, 2004.

Sriniwas Puri : *Soil Chemistry*, Oxford Book, Delhi, 2010.

T.V.S. Prasad : *Soil Chemistry : Nutrient and Water Management in Agricultural Soils*, Dominant Publication, Delhi, 2009.

Tapan Adhikari, S. Kundu and A. Subba Rao : *Nanotechnology in Soil Science and Plant Nutrition*, New India Publishing Agency, Delhi, 2013.

Index